Ecotourism Essentials

This comprehensive book, *Ecotourism Essentials: Principles, Challenges, and Practices for a Sustainable Future,* delves into the intricate dynamics of ecotourism sustainability and effective communication, offering insights from global analyses and scholarly research. Divided into three parts, this book covers a wide range of topics relevant to both researchers and practitioners.

This book navigates through three distinct parts. Part I illuminates the foundations of ecotourism sustainability and communication, examining how these principles shape the industry's ethos and operations. Part II delves into the challenges and critiques faced by the tourism industry, while Part III offers insights into ecotourism practices and perspectives, including discussions on animal-centered debates and the adoption of sustainability principles by leading ecolodges.

Ecotourism Essentials is an essential resource for anyone interested in understanding and promoting sustainable travel practices. Whether you're a researcher, practitioner, or simply passionate about ecotourism, this book offers valuable perspectives and insights into the evolving landscape of responsible travel.

The chapters in this book were originally published in *Tourism Recreation Research.*

Tzung-Cheng (TC) Huan is Professor at National Chiayi University, Taiwan. He is also Editor-in-Chief of *Tourism Recreation Research* and a member of the UNWTO (United Nations World Tourism Organization) Panel of Tourism Experts. He has been honored by the 2011–2012 Outstanding Reviewer Award from *Cornell Hospitality Quarterly*; the 2013, 2017, and 2018 Emerald Literati Awards; and the 2018 Publons Peer Review Award.

Aliana Man Wai Leong is Associate Vice-President for International Affairs and Dean, Professor, and Doctoral Supervisor at the School of Liberal Arts at the Macau University of Science and Technology (MUST), China. She is also former Dean of the Faculty of Hospitality and Tourism Management at MUST, China. She has been honored as one of the "Top Ten Most Influential Teachers in China's Tourism and Hospitality Education" and one of the "Top Ten Most Influential Teachers in National Tourism Institutions."

Tao Ma is Professor at the School of Management and Dean of the Business School at the Harbin Institute of Technology, China. His research interests include regional industrial planning and the economic analysis of resources and energy. He is recognized as a "Youth Talent" under the National Cultural Masters and Four Batches Talent Project and serves as Chief Expert for major projects funded by the National Social Science Fund of China.

Ecotourism Essentials

Principles, Challenges, and Practices for a Sustainable Future

Edited by
Tzung-Cheng (TC) Huan, Aliana Man Wai Leong and Tao Ma

Routledge
Taylor & Francis Group
LONDON AND NEW YORK

First published 2025
by Routledge
4 Park Square, Milton Park, Abingdon, Oxon, OX14 4RN

and by Routledge
605 Third Avenue, New York, NY 10158

Routledge is an imprint of the Taylor & Francis Group, an informa business

British Library Cataloguing-in-Publication Data
A catalogue record for this book is available from the British Library

ISBN13: 978-1-032-86266-8 (hbk)
ISBN13: 978-1-032-86270-5 (pbk)
ISBN13: 978-1-003-52211-9 (ebk)

DOI: 10.4324/9781003522119

Typeset in Myriad Pro
by codeMantra

Publisher's Note
The material in this volume has been reproduced in facsimile. This means we can retain the original pagination to facilitate easy and correct citation of the original essays. It also explains the variety of typefaces, page layouts and numbering.
The publisher accepts responsibility for any inconsistencies that may have arisen during the conversion of this book from journal articles to book chapters, namely the inclusion of journal terminology.

Contents

Citation Information

The following chapters were originally published in various issues and volumes of *Tourism Recreation Research*. When citing this material, please use the original page numbering for each article, as follows:

Chapter 1

Communicating sustainability and ecotourism principles by ecolodges: a global analysis
David A. Fennell and Danuta de Grosbois
Tourism Recreation Research volume 48, issue 3 (2023), pp. 333–351

Chapter 2

Sustainability for ecotourism: work identity and role of community capacity building
Clare D'Souza, Mehdi Taghian, Timothy Marjoribanks, Gillian Sullivan Mort, Md Manirujjaman and Stephen Singaraju
Tourism Recreation Research, volume 44, issue 4 (2019), pp. 533–549

Chapter 3

How eco-service quality affects tourist engagement behaviour in ecotourism: eco-consciousness and eco activity-based learning and S-O-R Theory
Indrani Paul and Gobinda Roy
Tourism Recreation Research
DOI - https://doi.org/10.1080/02508281.2023.2275240

Chapter 4

NGOs in ecotourism: patrons of sustainability or neo-colonial agents? Evidence from Africa
Amare Wondirad, Denis Tolkach and Brian King
Tourism Recreation Research, volume 45, issue 2 (2020), pp. 144–160

Chapter 5

Limits and paradoxes in the development of ecotourism in Kenya: implications on the sustainability of the natural environment
Bonface Odiara Kihima
Tourism Recreation Research, volume 41, issue 1 (2016), pp. 80–88

Chapter 6

Academia and the Evolution of Ecotourism
Bob Mckercher
Tourism Recreation Research, volume 35, issue 1 (2010), pp. 15–26

Chapter 7

Evolution of Ecotourism: No Crisis A Response to McKercher (2010)
Ralf Buckley
Tourism Recreation Research, volume 35, issue 3 (2010), pp. 311–313

Chapter 8

Ecotourism, Animals and Ecocentrism: A Re-examination of the Billfish Debate
David A. Fennell
Tourism Recreation Research, volume 38, issue 2 (2013), pp. 189–202

Chapter 9

Sustainability and ecotourism principles adoption by leading ecolodges: learning from best practices
Danuta de Grosbois and David A. Fennell
Tourism Recreation Research, volume 47, issue 5–6 (2022), pp. 483–498

Chapter 10

An Ecofeminist Approach to Ecotourism Development
Margaret Byrne Swain and Melissa Tandy Byrne Swain
Tourism Recreation Research, volume 29, issue 3 (2004), pp. 1–6

Chapter 11

Mass and Urban Ecotourism: New Manifestions of an Old Concept
David B. Weaver
Tourism Recreation Research, volume 30, issue 1 (2005), pp. 19–26

Chapter 12

Examining the Relationship Between Ecotourists and Philanthropic Behaviour
Martha L. Barnes and Paul Eagles
Tourism Recreation Research, volume 29, issue 3 (2004), pp. 35–38

Notes on Contributors

Martha L Barnes is Associate Professor in the Department of Recreation and Leisure Studies at Brock University, St. Catharines, Canada.

Ralf Buckley is former Director of the International Centre for Ecotourism Research and Emeritus Professor at Griffith University, Australia.

Danuta de Grosbois, PhD, is Associate Professor in the Department of Geography and Tourism Studies at Brock University, St. Catharines, Canada. Her areas of research interest include performance evaluation in tourism, sustainability reporting in the hospitality and tourism industry, and carbon footprint assessment and disclosure.

Clare D'Souza is Professor of Marketing (Sustainability) in the Department of Entrepreneurship, Innovation and Marketing at La Trobe Business School at College of Arts, Social Sciences & Commerce at La Trobe University, Melbourne, Australia.

Paul Eagles is Distinguished Professor Emeritus in the Department of Recreation and Leisure Studies at the University of Waterloo, Canada.

David A. Fennell researches mainly in the areas of ecotourism, tourism ethics, and moral issues tied to the use of animals in the tourism industry and sustainability. A major thrust of his research involves the use of theory from other disciplines (e.g., biology and philosophy) to gain traction on many of tourism's most persistent issues and problems. Fennell is founding Editor-in-Chief of the *Journal of Ecotourism*.

Tzung-Cheng (TC) Huan is Professor at National Chiayi University, Taiwan. He is also Editor-in-Chief of *Tourism Recreation Research* and a member of the UNWTO (United Nations World Tourism Organization) Panel of Tourism Experts. He has been honored by the 2011–2012 Outstanding Reviewer Award from *Cornell Hospitality Quarterly*; the 2013, 2017, and 2018 Emerald Literati Awards; and the 2018 Publons Peer Review Award.

Brian King is Professor and Associate Dean at the School of Hotel and Tourism Management at Hong Kong Polytechnic University, Hong Kong. His research interests include tourism, culture and communication, and hospitality and tourism leadership.

Aliana Man Wai Leong is Associate Vice-President for International Affairs and Dean, Professor, and Doctoral Supervisor at the School of Liberal Arts at the Macau University of Science and Technology (MUST), China. She is also former Dean of the Faculty of Hospitality and Tourism Management at MUST, China. She has been honored as one of the "Top Ten Most Influential Teachers in China's Tourism and Hospitality Education" and one of the "Top Ten Most Influential Teachers in National Tourism Institutions."

Tao Ma is Professor at the School of Management and Dean of the Business School at Harbin Institute of Technology, China. His research interests include regional industrial planning and the economic analysis of resources and energy. He is recognized as a "Youth Talent" under the National Cultural Masters and Four Batches Talent Project and serves as Chief Expert for major projects funded by the National Social Science Fund of China.

Md Manirujjaman is Adjunct Lecturer in Marketing in the Department of Entrepreneurship, Innovation and Marketing at La Trobe University, Melbourne, Australia.

Timothy Marjoribanks is Professor of Management and Acting Director at the Centre for Transformative Innovation at the School of Business, Law, and Entrepreneurship at Swinburne University, Melbourne, Australia.

Bob McKercher is Professor of Tourism at the School of Business at the University of Queensland, Brisbane, Australia.

Bonface Odiara Kihima is Associate Professor of Hospitality and Leisure Studies (DHLS) at the Technical University of Kenya, Nairobi, Kenya. While he focuses to come up with a research agenda that endeavors to promote discovery, creativity, and innovation, his main research interests are parks, natural areas, and ecotourism.

Indrani Paul is a doctoral student at International Management Institute, Kolkata, India. Her research interest includes the area of consumer engagement, ecotourism, and consumer psychology. She has published an article in the *Journal of Hospitality and Tourism Management*.

Gobinda Roy is Assistant Professor of Marketing at International Management Institute, Kolkata, India. Dr. Roy has earned his PhD from Vinod Gupta School of Management at Indian Institute of Technology, Kharagpur, India. He has over 20 years of experience in Industry and Academia. He has published articles in many international journals such as *International Journal of Hospitality Management, Tourism Recreation Research, Journal of Hospitality and Tourism Management, Journal of Retailing and Consumer Services,* and *Journal of Marketing communication*. His current research interest includes online/internet marketing, tourism marketing, consumer behavior, and online retails.

Stephen Singaraju is Head of Economics Programme Area and Senior Assistant Professor at Universiti Teknologi Brunei, Brunei.

Gillian Sullivan Mort is Dean at Eudemonia Institute, Brisbane, Australia.

Margaret Byrne Swain is former Co-Director of Women's Resources and Research Center at the University of California, Davis, USA.

Melissa Tandy Byrne Swain is a doctoral candidate and adjunct instructor at New York University, USA.

Mehdi Taghian is former Senior Lecturer in the Department of Marketing at Deakin Business School at Deakin University, Australia.

Denis Tolkach is Associate Professor at James Cook University, Australia. He has been teaching a variety of subjects in the hospitality and tourism field. His main interests are in tourism and development, community-based tourism, and nature-based tourism.

David B. Weaver is Professor of Tourism Management at the School of Hotel, Restaurant and Tourism Management at the University of South Carolina, Columbia, USA.

Amare Wondirad is Senior Lecturer at The University of the South Pacific, Suva, Fiji. He has completed his PhD at The Hong Kong Polytechnic University, School of Hotel and Tourism Management, Hong Kong. His research interests include ecotourism, sustainable tourism, and cruise tourism.

Introduction

Tzung-Cheng (TC) Huan, Aliana Man Wai Leong and Tao Ma

In an age where environmental concerns are becoming increasingly urgent, ecotourism stands as a pivotal concept at the intersection of travel and sustainability. *Ecotourism Essentials: Principles, Challenges, and Practices for a Sustainable Future* provides a comprehensive examination of this dynamic field, offering a critical exploration of how ecotourism principles are applied and communicated across various contexts.

Ecotourism has evolved from a niche market into a significant global industry that seeks to balance the enjoyment of travel with the imperative of environmental stewardship. This book delves into the intricate dynamics of ecotourism, illuminating both its foundational principles and its real-world applications. By bridging theory and practice, it serves as a crucial resource for anyone invested in the future of sustainable travel.

The structure of this book is designed to offer a thorough exploration of ecotourism through three distinct yet interrelated parts. Each section provides valuable insights that collectively enhance our understanding of ecotourism's impact and potential.

Part I: Ecotourism Sustainability and Communication establishes the groundwork by examining the core principles of ecotourism sustainability and the ways in which these principles are communicated. David A. Fennell and Danuta de Grosbois, in their chapter, explore how ecolodges around the world articulate their sustainability practices, offering a global analysis of effective communication strategies. The exploration continues with Clare D'Souza and colleagues, who address the role of community capacity building and the integration of sustainability within the work identity of ecotourism. This section culminates with a study by Indrani Paul and Gobinda Roy on how eco-service quality influences tourist engagement, drawing on eco-consciousness and activity-based learning to understand tourist behavior.

Part II: Challenges and Critiques in Ecotourism presents a critical perspective on the limitations and controversies within the field. Amare Wondirad, Denis Tolkach, and Brian King challenge the role of NGOs in ecotourism, questioning whether these organizations act as true patrons of sustainability or as agents of neo-colonialism. Bonface Odiara Kihima's examination of ecotourism in Kenya highlights the paradoxes and limits of ecotourism development, shedding light on its impact on the natural environment. The discussion extends to the academic evolution of ecotourism, with contributions from Bob McKercher and Ralf Buckley that address ongoing debates and responses within the scholarly community.

Part III: Ecotourism Practices and Perspectives offers practical insights and forward-looking perspectives on ecotourism practices. Davida Fennell's re-examination of the Billfish Debate provides a nuanced discussion on the ethical considerations of ecotourism involving animals. Danuta de Grosbois and David A. Fennell explore the adoption of sustainability principles by leading ecolodges, offering valuable lessons from best practices. The section also includes an innovative ecofeminist approach by Margaret Byrne Swain and Melissa Tandy Byrne Swain, which broadens the discourse on ecotourism development. David B. Weaver's analysis of mass and urban ecotourism presents new manifestations of traditional concepts, while Martha L. Barnes and Paul Eagles examine the relationship between ecotourists and philanthropic behavior, adding another layer to our understanding of ecotourism's impact.

Ecotourism Essentials is more than just an academic text; it is a reflection of the evolving landscape of responsible travel. As we face mounting environmental challenges, the need for sustainable travel practices becomes ever more pressing. This book provides a rich tapestry of perspectives that contribute to the ongoing dialogue on how to achieve balance between enjoyment and conservation.

We hope this volume will serve as an essential guide for researchers, practitioners, and enthusiasts alike, fostering a deeper understanding of the principles and practices that underpin successful ecotourism. By examining both the foundational theories and practical implementations, this book aims to inspire more thoughtful and effective approaches to sustainable travel.

In a world increasingly aware of its environmental footprint, *Ecotourism Essentials* stands as a beacon for those committed to making travel a force for good. We invite you to explore the chapters within and engage with the critical discussions that shape the future of ecotourism.

Part I

Ecotourism sustainability and communication

Communicating sustainability and ecotourism principles by ecolodges: a global analysis

David A. Fennell and Danuta de Grosbois

ABSTRACT
The purpose of this study was to address the need for a global-scale study of sustainability practices and ecotourism principles in ecolodges with a focus on their communication via the Internet. The study developed an assessment framework to evaluate ecotourism principles and sustainability communication in ecolodges across five dimensions (nature-based focus, the moral imperative, environmental sustainability, community development, and education and research), examined the state of communication of related initiatives by a sample of ecolodges from around a world, and identified areas of improvement for Internet communication. Results indicate high levels of communication across all five dimensions and presence of four distinct groups of ecolodges based on their approach to ecotourism principles communication: engaged communicators, selective communicators, average communicators and poor communicators.

Introduction

Ecotourism is a type of tourism that focuses on experiencing natural areas and highlights the importance of biodiversity conservation (Björk, 2000). It is often viewed as a means to simultaneously support the preservation of natural ecosystems and promote sustainable local development (Ross & Wall, 1999). Ecotourism development is prevalent in many destinations around the world. An essential part of this development is the growth of ecolodges, which provide accommodation for visitors engaging in ecotourism activities. Ecolodge operators constitute a unique group of small and medium enterprises (SME). The term ecolodge was introduced by the International Ecotourism Society (TIES) in 1994 to explicitly identify and distinguish lodges that are nature-depended and meet the philosophy and principles of ecotourism (Mehta, 2007). It is therefore expected that ecolodges follow ecotourism principles, which include a nature-based focus, local benefits and conservation (sustainable development), learning, and the ethical imperative (Fennell, 2020). The importance of sustainability to ecolodges has become manifest in practices that emphasize blending with the environment, recognizing the interdependence of human design with the natural world, relying on natural energy flows, using renewable energy sources, and eliminating waste (Anderson, 1993; Drumm et al., 2004; Hawkins et al., 1995; Mehta et al., 2002). At the same time, however, there are concerns that not all lodges follow sustainability principles (Mehta, 2007) and that the trend of building more extensive, internationally funded and managed facilities located in the sensitive natural resources may lead to the devastation of natural areas (Wight, 1997).

Ecotourists are increasingly interested in better understanding sustainability practices adopted by operators they stay with and being able to distinguish between true ecolodges and those that claim to be ecolodges but do not follow ecotourism principles. According to a study by Deng and Li (2015), ecotourists self-identify as more environmentally responsible, more dedicated to nature, supportive of accreditation programmes and likely to stay at ecolodges with sound environmental practices, even at a higher cost. Given the potential to attract these tourists, it is critical for ecolodges to credibly communicate to visitors their efforts to follow ecotourism principles and implement sustainability practices. In response to this need, ecolodge criteria have been incorporated into some ecotourism certification programmes (e.g. Green Globe, Certificate of Sustainable Tourism in Costa Rica, Eco-Rating Scheme in Kenya) with calls for international certification specifically focused on ecolodges (Mehta, 2007). Additionally, the Internet has been steadily growing as a popular and inexpensive way for companies to communicate their sustainability initiatives to the public and potential customers.

Internet communication of sustainability initiatives by ecolodges can be better understood within the context of broader research on corporate social responsibility (CSR) communication in general. CSR focuses on business responsibilities towards environment and society and can be defined as encompassing 'the economic, legal, ethical, and discretionary expectations that society has of organizations at a given point in time' (Carroll, 1979, p. 500).

CSR communication is emerging as an essential component of business strategies implemented by companies across many sectors of the economy. If done effectively, it can serve as a way to demonstrate a company's commitment to sustainability principles, create positive relationships with stakeholders, enhance reputation, and lead to higher employee motivation and lower operational risk. On the other hand, the lack of CSR communication may undermine a company's credibility and reduce positive perceptions by stakeholders (Du et al., 2010). Even if companies engage in CSR communication, the benefits are not guaranteed. The stakeholders may perceive poorly executed CSR communication as a way to mask fraudulent and greenwashing practices (Lyon and Maxwell, 2011). Nekhili et al. (2017) pointed out that a highly positive, over-reporting approach to CSR investment could be misunderstood as 'bragging' and lead to backlash effects and suspicion from stakeholders who may feel manipulated. In order to assist companies in designing their CSR reporting and communication strategies, many international organizations have been developing disclosure standards and guidelines. Although there are many tools to improve credibility and comprehensiveness of CSR disclosure, many of them are not always applicable in a SME context, which includes ecolodges. SMEs often face unique challenges related to CSR implementation and communication compared to larger companies, for example, a limited amount of funds, time or expertise (Vernon et al., 2003).

Despite the importance of ecolodges as a critical component of the ecotourism experience, empirical research on sustainability communication by ecolodges is limited. Ecolodges are included as part of several studies on the implementation of sustainability practices by eco-accommodation (Nelson, 2010) or by ecotourism operators (Donohoe & Needham, 2008; Massi & De Nisco, 2018). Only a few studies investigate the adoption of ecolodge sustainability or ecotourism principles (Grosbois & Fennell, 2021; National Geographic, 2019; Wallace & Pierce, 1996) and their communication using the Internet (Lai & Shafer, 2005; Millar & Sammons, 2006; Sangpikul, 2010). Each of these studies focuses on a specific geographical area only. Consequently, no study focuses on the extent of sustainability practices communication by ecolodges around the world. A recent study by National Geographic investigated the adoption of sustainability practices by the members of the Unique Lodges of the World (National Geographic, 2019). However, since the data were collected by surveying ecolodge owners and managers, findings do not provide insight into the nature of the communication of these practices to external stakeholders.

The purpose of the current study is to address the need for a global-scale study of sustainability practices and ecotourism principles in ecolodges with a focus on their communication (via websites) to potential guests and other stakeholders. To meet this purpose, this study develops an assessment framework to evaluate ecotourism principles and sustainability communication in ecolodges, examines the state of communication of these initiatives by a sample of ecolodges from around the world, and identifies areas of improvement for Internet communication. This paper contributes to the literature in several unique ways. First, through its comprehensiveness, since it investigates five dimensions and 20 sub-dimensions of ecolodge sustainability. Second, unlike the previous studies that have limited sample sizes or are geographical area-specific, the current paper uses a larger sample, consisting of a cross-section of the world's leading ecolodges, allowing for a richer understanding of the state of CSR communication in the industry.

Literature review

Adoption and communication of sustainability practices

Despite the limited research on sustainability communication at ecolodges, valuable insight into this topic can be gained from the literature on ecotourism and the literature on corporate social responsibility (CSR) communication, especially in the context of SMEs. CSR focuses on business responsibilities towards environment and society and can be defined as encompassing 'the economic, legal, ethical, and discretionary expectations that society has of organizations at a given point in time' (Carroll, 1979, p. 500). CSR can be viewed instrumentally as a firm's behaviour that can affect its financial results or as a goal adopted by a firm independently of the financial considerations. As customers increasingly expect companies to embed sustainability values in their products, services and actions (Lii & Lee, 2012), interest in CSR has also been growing in disciplines such as management, economics and sociology. While several theories have been developed and applied to CSR research, the present study focuses on stakeholder and institutional legitimacy theories. Stakeholder theory stresses the obligations and responsibilities of a firm not only to address its own goals but also the goals and needs of its stakeholders, defined as individuals or groups that can affect or are affected by the organization's decisions (Freeman, 1984). Proponents of this theory argue that organizations need to engage different stakeholders in a collaborative decision-making process which considers the needs of multiple parties. The choice of stakeholder issues to address is related to stakeholder 'salience', defined by Mitchell et al. (1997) as 'the degree to which managers give priority to competing stakeholder claims', based on power, urgency and legitimacy' (p. 854). Stakeholder theory is divided into descriptive, normative and instrumental streams (Donaldson & Preston, 1995) that help practitioners and researchers understand relationships between firms and their stakeholders. The core hypothesis of instrumental stakeholder theory states that developing highly ethical stakeholder relationships characterized by fairness, trustworthiness, cooperation, care, respect and information sharing will lead to competitive advantage and improved financial performance of the operator (Hendry, 2001; Jones et al., 2018). In the context of ecolodges, the premise of building ethical relationships with stakeholders is especially relevant, as it constitutes a core ecotourism principle. Further research on stakeholder theory in the context of SMEs points out that stakeholder relationships in small businesses are often informal (Jenkins, 2004) and that relative importance of stakeholders and their concerns may depend on stakeholder social proximity to the operator (Lähdesmäki et al., 2019).

CSR reporting is defined as the 'process of communicating the social and environmental effects of organizations' economic actions to particular groups within society and to society at large' (Gray et al., 1996, p. 3), with a considerable emphasis in this research on understanding the motivation and choices that companies make about communication efforts. According to institutional legitimacy theory, firms adopt specific behaviours to obtain support from critical stakeholders and to gain access to resources (DiMaggio & Powell, 1983; Meyer & Rowan, 1977; Oliver, 1991; Scott, 2004), and therefore they choose to disclose CSR in response to growing public concern over environmental and social impacts of business operations. Disclosing CSR information allows companies to obtain social license to operate and gain legitimacy, defined as a 'generalized perception or assumption that the actions of an entity are desirable, proper, or appropriate within some socially constructed system of norms, values, beliefs, and definitions' (Suchman, 1995, p. 574). Legitimacy theory further proposes that information may be disclosed selectively, where only the positive content is provided, but adverse outcomes are not reported (Elsbach & Sutton, 1992; Meyer & Rowan, 1977). Several authors argue that stakeholder engagement plays a vital role in companies' decisions regarding relevance and materiality of CSR issues and therefore affects the final choice of information and data to be included in CSR communications (Gray, 2000).

The importance of CSR communication is widely recognized by both international organizations and the research community. The literature suggests that effective communication of CSR has both external and internal benefits. It can lead to a company being perceived as socially responsible and create benefits such as greater customer loyalty, higher levels of customer satisfaction, improved company reputation, better efficiency, improved dialogue with external stakeholders, avoidance of stakeholders' criticism, and better relations with the local community (Isenmann, 2006; Kim & Park, 2011; Sen & Bhattacharya, 2001). Adoption and disclosure of CSR can further become a form of reputation insurance, generating moral capital and goodwill by enhancing the company's image and reputation among stakeholders, which in turn can serve as insurance protecting the company in the aftermath of adverse events (Godfrey et al., 2009). Customers are also likely to reward companies that engage in and disclose CSR activities by purchasing more of their products and services and giving them higher evaluations and positive feedback (Sen & Bhattacharya, 2001). In the context of ecolodges, the ecotourists have strong environmental consciousness and social caring, making these relationships even more potent.

While CSR communication can be done in many different ways, such as through workshops, annual reports, news articles, public speeches, or advertisements, the Internet is increasingly recognized as a powerful tool that enables companies to present their activities at an international scale. Research shows that although larger companies' annual reports are one of the essential tools of CSR disclosure, smaller companies consider websites to be a more convenient way to provide information to their stakeholders. The Internet is viewed as an inexpensive, fast and practical information dissemination tool that can reach many stakeholders simultaneously (1998; Kent et al., 2003). A company's website can fulfil several functions, such as educate or inform customers, build and maintain customer relationships, reduce uncertainty, play an image-creating role (Sullivan, 1999) or enable two-way communication and obtain feedback from stakeholders

(Pollach, 2005). Previous research has demonstrated the growing importance of the Internet and the corporate websites as communication tools (Hill & White, 2000), especially for CSR and sustainability-related efforts (Esrock & Leichty, 1998, 2000). However, despite the recognition that initiatives undertaken by a company to gain legitimacy and trust need to be communicated to stakeholders, some authors point out that not all companies communicate their sustainability efforts effectively, even if they are committed to fulfilling their social responsibilities (Dawkins, 2005; Lewis, 2003). For ecolodges, which operate within the context of ecotourism, sustainability constitutes the core element of their product, and effective communication of sustainability efforts and commitment to ecotourism principles are needed to ensure legitimacy.

Companies can use their websites to communicate with stakeholders in two ways: unidirectional and bidirectional (Esrock & Leichty, 1998; White & Raman, 1999). Unidirectional communication involves the dissemination of information with little or no interactivity; while bidirectional includes dialogue and interaction between the company and its stakeholders along with the willingness to respond to concerns raised by the public. Studies on CSR communication using the Internet, specifically, focus primarily on the topics presented, stakeholders, and factors that affect the amount and type of disclosure that is made (Holder-Webb et al., 2009; Hou & Reber, 2011). Research on CSR communications in the context of SME, such as ecolodge operators, recognizes significant differences between SMEs and larger companies in terms of their strategy, implementation of new practices, and approach to sustainability (Cromie et al., 1995; Jenkins, 2006). Distinct characteristics of SMEs include a smaller number of employees, lower annual turnover, independence, resource constraints, reliance on personal relationships and informality, active management by owners, local operations and dependence on internal sources to finance growth (McAdam et al., 2014; Soundararajan et al., 2018). According to some authors, many of these distinct characteristics of SMEs facilitate the internal implementation of CSR practices, but at the same time limit their external communication and reporting (Baumann-Pauly et al., 2013). Therefore, it is recognized that family SMEs may engage informally in CSR practices without explicitly using the CSR label or engaging in CSR communications (Baumann-Pauly et al., 2013; Fitzgerald et al., 2010). CSR implementation in SMEs may start with owners and managers addressing social and environmental concerns, often under the influence of stakeholders (Pedersen, 2006), in response to changing regulations or based on the personal values, beliefs, and background of the founders (Fitzgerald et al., 2010). However, lack of formal communication of these initiatives may negatively affect reputation and stakeholder relations for these SMEs (Nielsen & Thomsen, 2009).

Ecolodges and CSR

Ecolodges are an integral part of the infrastructure needed to develop and provide ecotourism services and as such, are expected to incorporate ecotourism principles into their operations. Wallace and Pierce (1996) listed six principles that an ecolodge should follow: minimize negative impacts on the environment and local people, raise visitors' awareness of understanding of natural and cultural systems in the area visited, contribute to the conservation and management of protected areas, ensure participation of local people in ecotourism development, share economic and other benefits with local people without overwhelming or replacing traditional practices, and create opportunities for local people to visit natural areas. Ceballos-Lascuráin (1996) stressed ecolodge design features that should blend with their surroundings. Similarly, The International Finance Corporation (2004) stated that ecolodges must incorporate three main components: conservation of neighbouring lands, benefits to local communities, and interpretation to both local populations and guests. A more recent definition by Mehta (2007) specified that an ecolodge is 'a 5–75-room, low-impact, nature-based, financially sustainable accommodation facility' that meets the following objectives: 'helps protect sensitive neighbouring areas; involves and benefits local communities; offers tourists an interpretive and interactive participatory experience; provides a spiritual communion with nature and culture; and is planned, designed, constructed and operated in an environmentally and socially sensitive manner' (Mehta, 2007, p. 419). Osland and Mackoy (2004) further distinguished between dedicated, casual, scientific and agri-ecolodges based on the lodge setting and the type of activities that visitors engage in while they stay at the lodge.

There is little doubt that ecolodges are an essential part of ecotourism and influence the sustainability of ecotourism offerings. From the stakeholder perspective, operations of ecolodges can influence owners and employees through direct management practices, local community through their employment and procurement practices, tourists through the choice of activities and education offered, and the environment through design and ongoing management. If they are poorly designed, built or operated, ecolodges have the potential to harm different stakeholder groups, especially the environment and local communities (Millar & Sammons, 2006; Osland & Mackoy, 2004). According to Russell et al. (1995) lodges that in practice do not meet the philosophy or principles of ecotourism, even if they call themselves ecolodges, are not indeed ecolodges. However, as there are no internationally accepted guidelines or certifications for ecolodges, there is a significant diversity and confusion in the eco-lodge market, where nature-based tourist lodges, such as fishing and ski lodges may also be called ecolodges (Fennell & De Grosbois, 2017). It is therefore crucial for ecolodges to ensure their customers and the public are aware of their commitment to ecotourism principles and sustainability.

While some of the world's most recognized green certification programmes include criteria for lodges and offer an opportunity to get a third-party assurance of ecotourism principles adoption, many operators, especially small ones, find these costs-prohibitive. In such situations, communicating ecotourism and sustainability practices directly to stakeholders is increasingly done through other means, especially the Internet. Previous studies have demonstrated that ecotourists use the Internet as an effective method for gathering travel and tour information when planning ecotourism activities (Donohoe & Needham, 2008; Lai & Shafer, 2005). However, despite the importance of the Internet from the tourists' perspective, the previous research points to limited communication of ecotourism principles and sustainability initiatives by ecotourism operators and ecolodges. In their study of Internet use by 35 ecolodges in Latin America and the Caribbean, Lai and Shafer (2005) found that online ecolodge communications did not completely align with the main principles of ecotourism and some areas, such as education, were underrepresented. Donohoe and Needham (2008) and Sangpikul (2010), who studied ecotourism providers, including ecolodges, in Canada and Thailand respectively, also found limited alignment between their messages on the Internet and ecotourism principles. Similarly, in a study of websites of 53 ecolodges in Costa Rica, Millar and Sammons (2006) concluded that ecolodges failed to fully utilize the Internet to bring their product to the market. All the previous studies included a limited number of ecotourism principles and focused on only one geographical area each, therefore leading to a lack of a global perspective on sustainability reporting by ecolodges.

Methodology

This section discusses the characteristics of the sample, the content analysis approach used in the study and the development of an ecolodge assessment framework. The final sub-section discusses the use of cluster analysis to better understand patterns of ecotourism communications in the sample.

Sample

There is no one source to identify the global population of ecolodges to draw a representative sample. The large number of self-claimed ecolodges identified using the Internet range considerably based on the nature of their operations and may include lodges that do not adopt ecotourism principles. Previous empirical studies faced similar challenges. Lai and Shafer (2005) called any efforts to estimate the size of ecotour operators' population 'problematic and controversial'. Donohoe and Needham (2008) identified the lack of a comprehensive inventory of Canadian ecotourism providers as a barrier to the selection of a statistically representative sample.

Since the purpose of this study was to provide an exploratory understanding of the current state of sustainability communications by ecolodges, the sample needed to include the true ecolodges that were most likely to follow ecotourism and sustainability principles and could serve as a source of information on the extent and nature of CSR communications. Therefore, the sample was purposefully selected based not on ecolodge self-declarations, but independent third party ecolodge designations. Therefore, the sample included:

(1) lodges listed in the *Authentic Ecolodges* book written by H. Mehta (2010)
(2) lodges that were recipients of the World Travel Awards in the ecolodge category during years 2007–2019
(3) lodges listed by the National Geographic on the 50 top ecolodges list (Christ & Siber, 2008)

Content analysis

In order to provide an assessment of sustainability and ecotourism principles' communication in the ecolodge sector, this study used content analysis (Stemler, 2000) of websites of leading ecolodges located around the world. The data analysis involved both qualitative and quantitative content analysis. The data collected included all content related to ecotourism and sustainability principles and initiatives reported on ecolodge websites in the form of different documents and information directly displayed in the website's links, but excluding information in the form of videos and links to external news articles. If an ecolodge was a part of a group and had a link to its parent company's website or owner's website, the information on these websites was also included if it applied to the specific ecolodge under consideration. Finally, if its foundation ran sustainability-related efforts of a given ecolodge, the information provided on the foundation website was included if it was linked from the ecolodge's website. In order to ensure comparability of the provided information and to take into consideration the potential for the website content to change over time, all the websites were coded within a short period of time (May to June 2020).

In accordance with previous research (Jose & Lee, 2007; Stemler, 2000) we developed an *a priori* coding scheme based on the four ecotourism principles, as proposed in the ecotourism framework developed by Fennell (2020). The coding scheme was further modified and expanded using qualitative content analysis of data collected for this study. Qualitative content analysis and emergent coding (Stemler, 2000) were applied to organize the data into themes, sub-themes and practices within each ecotourism principle. These categories were identified through the review and coding process which involved adding new codes, removing other codes, as well as renaming and merging themes and groups of practices through the use of open, axial, and selective coding. The categories identified through this process became the dimensions, sub-dimensions and indicators in the ecolodge sustainability assessment framework used to evaluate ecolodge communications. The final framework contained five dimensions and 20 sub-dimensions with a total of 79 indicators as outlined in more detail in Table 2. All the indicators are listed in Table 3.

Once the assessment framework was finalized, the data was analyzed again using quantitative content analysis. Quantitative content analysis is used to investigate whether specific words or concepts are present in the text. It has been used previously both in corporate social responsibility research (Gray et al., 1995) and in the ecotourism research (Donohoe & Needham, 2008; Lai & Shafer, 2005). All the written content on each website was analyzed using occurrence-based content analysis (Carley, 1993) to identify whether the indicators from each dimension and sub-dimension of ecotourism were present or not and the results of the analysis were recorded. The websites represented the sampling units. In order to determine whether a specific dimension or a criterion was present or not, we investigated cues, i.e. specific initiatives mentioned on the websites. We identified them as belonging to one of the categories in the ecotourism assessment framework developed for this study. (Table 1).

Cluster analysis

A cluster analysis was conducted to better understand different patterns of ecotourism communications in the sample. Clustering is used to group objects that are similar to each other, but different from objects belonging to other groups (Bramer, 2020) and has been used to assess CSR disclosures in previous research (Alberici & Querci, 2016). For each of the five ecotourism dimensions, the percentage of indicators communicated by each lodge was calculated, resulting in the following variables: percentage of nature-based indicators present, percentage of moral imperative indicators present, percentage of environment and conservation indicators present, percentage of local community development indicators present, and percentage of education and research indicators present. The clustering approach used was the K-means method with Euclidean

Table 1. Ecotourism dimensions and sub-dimensions for ecolodges.

NATURE-BASED
(1) Location in natural settings
(2) Facilities (accommodation and infrastructure) with good access to nature
(3) Availability of nature-related activities
MORAL IMPERATIVE
(1) Explicit statement indicating a commitment to sustainability or ecotourism
(2) External recognition for sustainability/ecotourism efforts
(3) Accountability
(4) Sharing profits with the community
(5) Contributions to social justice
SUSTAINABILITY 1: ENVIRONMENT AND CONSERVATION
(1) Sustainable lodge design and construction
(2) Environmentally-friendly lodge operations
(2.1) Water management
(2.2) Energy management
(2.3) Waste and supply chain management
(2.4) Pollution and carbon footprint reduction
(3) Biodiversity conservation
SUSTAINABILITY 2: LOCAL COMMUNITY DEVELOPMENT
(1) Preserving socio-cultural heritage
(2) Social wellbeing
(2.1) Supporting education
(2.2) Supporting health
(2.3) Supporting quality of life
(2.4) Employee wellbeing
(3) Contribution to the local economy
(4) Involving local people in ecolodge planning, design, construction and operations
(5) Minimizing negative social impacts
EDUCATION AND RESEARCH
(1) Learning opportunities for tourists
(1.1) Environment-focused
(1.2) Local culture-focused
(2) Influencing local people and staff attitudes to conservation and ecotourism
(3) Opportunities for tourists to get involved
(4) Supporting research

distance. The final number of clusters was inferred based on the number of cases in each cluster along with final centroids, which are the average values of the objects included in each cluster. Interpreting the clusters included the analysis of centroids and descriptive variables for ecolodges in each cluster.

Results

Once the entries that were the same were removed, the resulting sample consisted of 83 ecolodges. Additionally, five lodges with no websites and 13 lodges that were tent-based were eliminated from the sample, resulting in the final sample size of 65 properties. The ecolodges that were included in the final sample are listed in Table 2. The majority of them (67.7%) were operated as single properties, with owners not owning or managing other lodges or as single conservation areas, whose owners could operate several lodges in the same reserve but did not own or operate any properties beyond that reserve. The remaining lodges were each part of a larger group of properties or were owned by a larger company such as a tour operator. The lodges were located mostly in Africa (33.8% of the sample), Central and South America (21.5%) and Asia (20.0%). The remaining lodges were in Oceania, North America, the Caribbean, the Middle East and Europe.

Ecotourism principles communication

The majority of ecolodges in the sample (70.8%) addressed all five dimensions of ecotourism. The most popular dimension was 'nature-based', which was met by all the ecolodges in the sample. 'Environmental sustainability', 'community development' and 'education and research' were all addressed by over 90% of the ecolodges each. The only dimension with lower communication level was 'moral imperative' which was addressed by 81.5% of the ecolodges. (Figure 1).

Nature-based

All three criteria within the 'nature-based' dimension were communicated by 100% of ecolodges. All lodges in the sample explained on their websites that they are located in natural settings, often within or close to a public or private reserve or conservation area. The most popular locations were rainforest, jungle and temperate forest (27%) and savanna (25%). The remaining lodges were located in water-based locations with beaches, including islands, bays or rivers (21%), mountains (18%) and

Table 2. List of ecolodges included in the final sample.

Ecolodge Name, Location	Ecolodge Name, Location
Adrère Amellal Eco Lodge, Egypt	Kumbuk River, Sri Lanka
Al Maha Desert Resort & Spa, United Arab Emirates	Lapa Rios Lodge, Costa Rica
Apani Dhani Ecolodge, India	Manda Wilderness Nkwichi Lodge, Mozambique
Aristi Mountain Resort & Villas, Greece	Mashpi Lodge, Ecuador
Basata Ecolodge, Egypt	Milia Mountain Retreat, Greece
Bay of Fires Lodge, Australia	Morgan's Rock Hacienda & Ecolodge, Nicaragua
Black Sheep Inn Ecolodge, Ecuador	Mosetiha Bush Camp, South Africa
Borana Lodge, Kenya	NAPO Wildlife Center, Ecuador
Bushmans Kloof Wilderness Reserve & Wellness Retreat, South Africa	Nihi Sumba (formerly Nihiwatu), Indonesia
Caiman Ecological Refuge, Brazil	North Island, Seychelles
Campi ya Kanzi, Kenya	O'Reilly's Rainforest Retreat, Australia
Canopy Tower, Panama	Phinda Forest Lodge, South Africa
Chalalàn Ecolodge, Bolivia	Posada Amazonas Lodge, Peru
Chumbe Island Coral Park, Tanzania	Ranweli Holiday Village, Sri Lanka
Coconut Lagoon, India	Rosalie Forest Eco Lodge, Dominica
Cree Village Eco Lodge, Canada	Sanctuary Olonana, Kenya
Cristalino Jungle Lodge, Brazil	Sarova Shaba Game Lodge, Kenya
Crosswaters Ecolodge & Spa, China	Shamwari Game Reserve – Eagles Crag Lodge, South Africa
Daintree Eco Lodge & Spa, Australia	Six Senses Hideaway Ninh Van Bay, Vietnam
Ecolodge Rendez-Vous, Saba Island	Soneva Fushi, Maldives
El Monte Sagrado, New Mexico, USA	Sukau Rainforest Lodge, Malaysia
Feynan Ecolodge, Jordan	The Boat Landing Guest House, Laos
Finca Esperanza Verde Ecolodge, Nicaragua	The Lodge at Chaa Creek, Belize
Highland Center, New Hampshire, USA	Tiamo Resort, Bahamas
Il Ngwesi Eco-Lodge, Kenya	Tiger Mountain Pokhara Lodge, Nepal
Jock Safari Lodge, South Africa	Topas Ecolodge, Vietnam
Kapawi Ecolodge, Ecuador	Tourindia Kettuvaliarn Houseboats, India
Kasbah Du Toubkal, Morocco	Tswalu Kalahari Reserve, South Africa
Kaya Mawa, Malawi	Uruyén Lodge, Venezuela
Kicking Horse River Lodge, Canada	Vamizi Island Resort, Mozambique
Kleins Camp, Tanzania	Vumbura Plains Camp, Botswana
Knapdale Ecolodge, New Zealand	Wilderness Lodge Arthur's Pass, New Zealand
	Wolwedans, Namibia

Note: Adapted from Grosbois & Fennell, 2021.

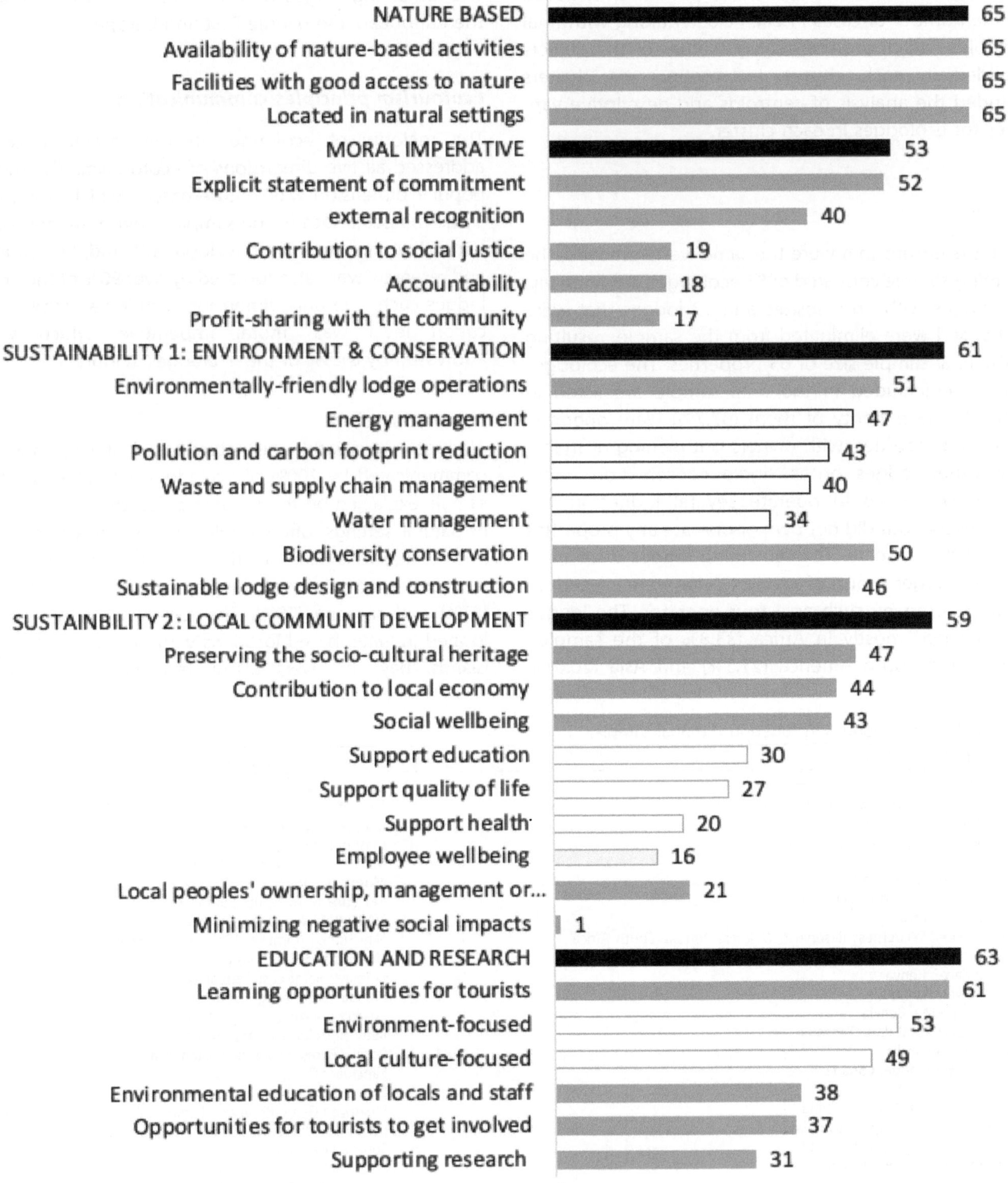

Figure 1. Number of lodges addressing ecotourism dimensions and sub-dimensions.

deserts (9%). All ecolodges also posted information about their facilities providing reasonable access to nature and listed possible nature-based activities tourists could engage in while visiting. These activities included both land-based activities and water-based activities, such as hiking, night walks, nature walks, visiting protected areas, bird watching, horseback riding, wildlife watching, safaris, star watching, surfing, fishing, canoeing, and snorkelling.

Moral imperative

Moral imperative communications were assessed by looking at five criteria. The criterion addressed by the largest number of lodges (80.0%) was an explicit statement on the website, indicating a commitment to sustainability or ecotourism. The statements provided by the lodges ranged from the brief announcement of the importance of sustainability to mission statements, our philosophy statements, declarations of guiding values, environmental policies, and environmental management plans. The second criterion used was mentioning an external recognition for ecotourism or sustainability efforts and was addressed by 61.5% of lodges. The forms of recognition communicated by the lodges included mostly third-party sustainability or environmental awards, certifications and external audits. The remaining three criteria were addressed by less than one-third of the lodges in the sample.

Contributions to social justice, communicated by 29.2% of lodges, focused mostly on gender equality through facilitation and promotion of women's economic activities, although other causes such as supporting refugees or improving access to the market for farmers were also present. The contributions to gender equality had three primary forms: education for women, employment opportunities for women and supporting women-run social enterprises. Regarding education, lodges offered training for women to produce and sell sanitary packs, become camp guides or members of the security team. Examples of additional employment opportunities for women included running cooking classes or creating beadwork for a jewellery brand managed by the lodge. The last type of gender equality effort was support for social enterprises, such as laundry services provision by local women, beach baskets for guests made by vulnerable single mothers, or women's cooperatives where local women make pottery, candles, towels or other items. Contributing to social justice was also expressed by explicitly declaring a commitment to equality and elimination of discrimination against gender, race, religion or sexual orientation.

The accountability criterion reflected sustainability performance reporting through an annual report, a newsletter or a website section dedicated to performance reporting. It is important to note that although 27.7% of lodges provided performance reporting, some of them provided undated or outdated reports; for one of the lodges, the only report available was from 2010. The criterion of sharing profits with the local community was met by 26.2% of lodges and applied to situations when the community ran the lodge. Therefore all its revenues were reinvested into community projects or when the lodge committed to a fixed amount or percentage of revenues as a contribution to the community. For example, some lodges collected bed-night levies, a dedicated portion of each booking to promote development in the area or ensure that all revenues from guests visits to a local village go directly to the visited community.

Environmental sustainability

In order to assess environmental sustainability communications, three sub-dimensions were considered: sustainable and low-impact lodge design and construction, environmentally friendly lodge operations and contribution to biodiversity conservation. The criteria and indicators identified in each sub-dimension are presented in Figure 2. The data indicate that 70.8% of the lodges communicated information about their efforts towards sustainable lodge design and construction. The main themes identified in this sub-dimension, in descending popularity, were: using local materials in construction (41.5%), integrating the lodge with nature (33.8%), using environmentally friendly or sustainable materials (26.2%), using sustainable construction and energy-efficient design (18.5%), and taking into account the environment when choosing the location (4.6%).

The sub-dimension of environmentally friendly lodge operations was addressed by 78.5% of the lodges. These efforts were grouped under four categories: water management (addressed by 52.3% of lodges), energy management (72.3%), waste and supply chain management (61.5%), and pollution and carbon footprint reduction (66.2%). For each category, central themes were identified through qualitative content analysis. The themes related to water management efforts included: water harvesting and storage, water conservation and dealing with wastewater. Energy management-related themes included energy conservation, use of renewable energy to power the lodge and use of solar water heating. Concerning energy conservation, some lodges focused on implementing initiatives that reduce energy use, while others eliminated electrical lighting and used candles or solar-powered lighting instead. The lodges also differed in their approach to air conditioning, with some trying to make it more energy-efficient by using a heat pump or limiting hours of operations, while others chose not to provide air conditioning in guest rooms at all. Solar water heating was the most popular, and in some cases the only, use of renewable energy at lodges in the sample. Analysis of waste management and supply chain-related efforts led to the identification of six themes: waste reduction; waste sorting, reusing and recycling; organic waste disposal; sewage treatment; buying from sustainable suppliers; and buying sustainable food. Among the waste reduction efforts, the most popular

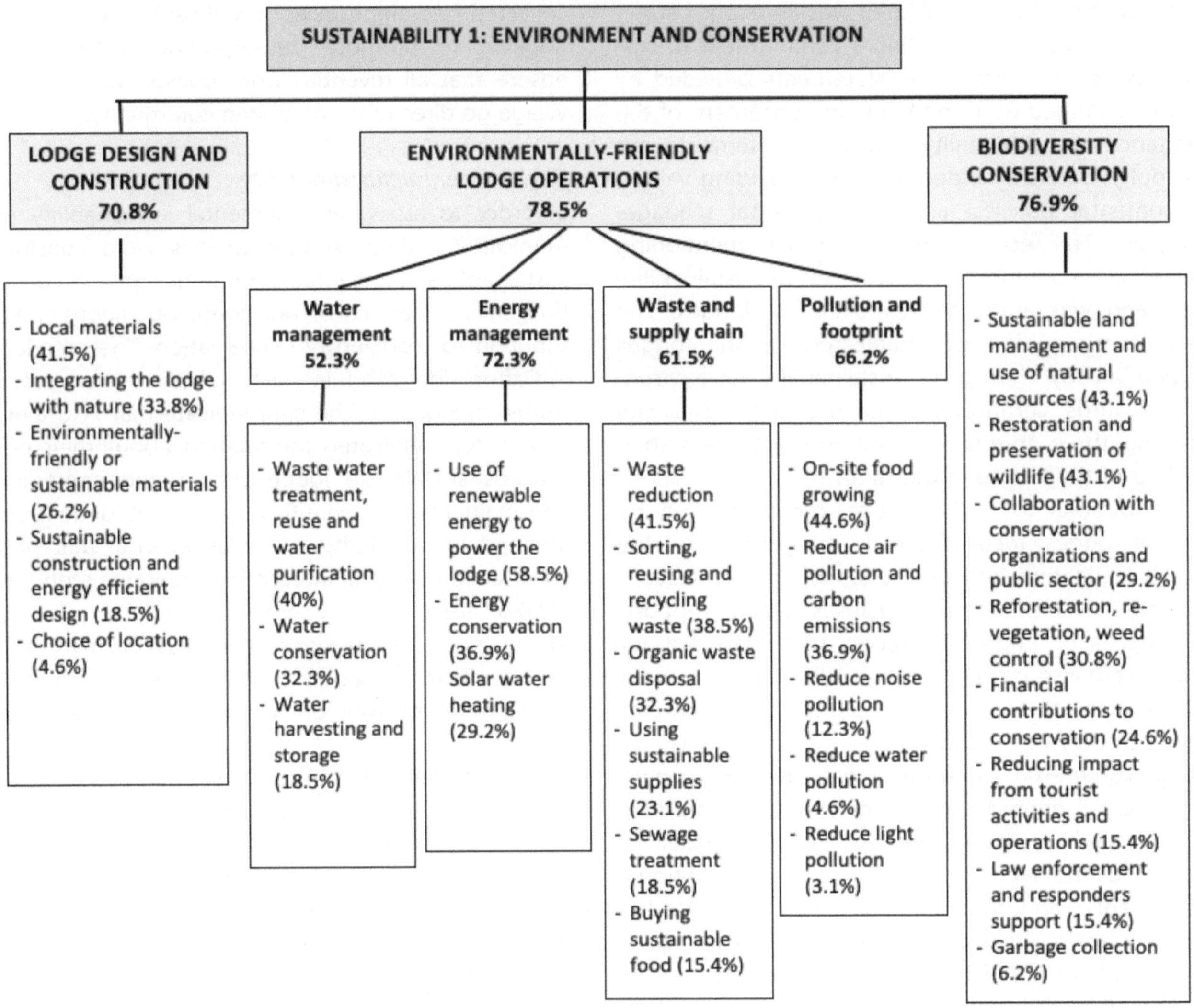

Figure 2. Environmental sustainability sub-dimensions and indicators and percentage of lodges reporting on them.

was the elimination of plastic bottles. Lodges that committed to the elimination of plastic bottles provided their guests with alternatives such as stainless steel water bottles, ceramic cups, locally produced clay jars as well as complimentary water filling stations with filtered water.

Contributions to the last sub-dimension of environmental sustainability, 'biodiversity conservation' was reported by 76.9% of the lodges and included many different initiatives, which were grouped into eight themes. The most popular theme was providing a financial contribution to conservation causes in the form of donations. In addition to financial contributions, lodges focused on collaboration with conservation organizations and the public sector to ensure sustainable development and biodiversity protection. The third theme, sustainable land management and use of natural resources encompass holistic efforts aimed at preserving a specific area. Several lodge owners who also own land, designated their land as private conservation areas or private reserves to protect the land and limit extractive activities. In some cases, lodge owners purchased large farms and worked on restoring them to their natural state. Sustainable land management initiatives included habitat management projects, implementation of reserve management plans, ongoing monitoring of water quality, rainfall, vegetation cover and wildlife.

The next two themes focused on the protection and restoration of vegetation and wildlife, respectively. Among vegetation related initiatives, the most popular were reforestation, planting indigenous plants and weed control. For the preservation of wildlife, the most popular was the reintroduction of animals and animal rescue. Depending on location, in order to protect wildlife, several lodges reimbursed local people for their livestock killed by predators, so they would not kill wildlife. Themes with lower popularity included law enforcement and responders' support, garbage collection in natural areas and reducing the impact of tourist activities and operations.

Community development

Ecolodges contribute to local community development in many ways, captured by five sub-dimensions in the framework. The most popular sub-dimension was preserving the socio-cultural heritage, addressed by 72.3% of lodges. Four main themes were discovered within this sub-dimension: using traditional architecture and techniques in lodge construction, incorporating local art, customs and traditions in the lodge and its operations, serving native dishes at the lodge and educating locals about the culture and heritage to help preserve it. Only 15.4% of the lodges addressed three or more of these themes, while 30.7% of them addressed one theme only. The most-reported theme was using traditional architecture and techniques, followed by serving native dishes.

The second most popular sub-dimension of local community development was a contribution to the economy. Lodges contribute to economic wellbeing mostly through hiring locals. Several lodges in the sample reported that 100% of their staff came from local communities. Lodges can also contribute to the economy by purchasing goods, services and food locally. Local purchases create demand for local goods and support many small producers and merchants. Lodges were also able to create opportunities for local businesses to sell their goods at the lodge or bring visitors to communities for shopping. Finally, lodges in the sample assisted local people in setting up their own business by providing training, advice or resources.

The sub-dimension of social wellbeing consisted of supporting education, supporting health, supporting other aspects of quality of life and ensuring employee wellbeing. Education was the most critical aspect for the lodges, with almost half of them (46.2%) addressing this issue. Education-related efforts not only included support for local schools in terms of infrastructure, teachers and resources but also extended beyond local schools to support youth to continue education and find a stable career. At the same time, less than a quarter of the lodges addressed the issue of employee wellbeing. The issues that were addressed in this theme included working condition, salaries, training and benefits such as paid leave to volunteer for charities.

In terms of involving local people, two main themes were present: involvement of the local or indigenous community in design and construction and local or indigenous ownership and management. Using local craftsmen, artisans, architects, or other specialists was the most popular case. Lodge ownership and management by local people was much less frequently cited. (Figure 3).

Education and research

Almost all (96.9%) lodges addressed the issue of education and research, mostly through offering learning opportunities for visitors (Figure 4). The most popular form of education for visitors was environment-based (81.5%), especially education about nature and science, available at 70.8% of lodges and taking place mostly through guided nature and wildlife tours and educational materials at the lodge. Significantly fewer lodges educated visitors about environmental sustainability or lodge's sustainability-related efforts (other than through the website).

Over three quarters (75.4%) of the lodges provided education about local culture. The main form of these efforts was providing tours and visits to local cultural and heritage attractions and places. 36.9% of ecolodges, offered ecotourists an opportunity to engage in authentic cultural experiences with local or indigenous people, which involved mostly community-led workshops or homestays.

Education efforts undertaken by the lodges did not focus only on the visitors but also extended to the locals and employees of the lodge. The most popular theme that emerged was 'influencing local people and staff attitudes to conservation and ecotourism' which involved providing environmental education for the local community to raise environmental awareness and to promote appreciation of nature, wildlife and conservation in order to change consumptive and unsustainable activities of local people. A large number of these efforts were directed at school children. Efforts aimed at the larger community involved community stakeholder workshops and events during which environmental education was offered to teachers, fishers, farmers, other community groups and government officials free of charge. During workshops, local participants learn about the importance of the ecosystem and discuss related topics, such as waste, fire, poaching, rainfall, prospecting and conservation, sustainable living, renewable energy, permaculture, impacts of ecolodge operations, reducing, reusing and recycling or other environmental practices.

The second theme was demonstrating the value of conservation and ecotourism as an alternative economic activity and employment for the local population. Lodges in the sample promoted the idea that ecotourism and nature tourism can provide more opportunities and often-higher benefits for locals compared to resource exploitation. It can also create different career

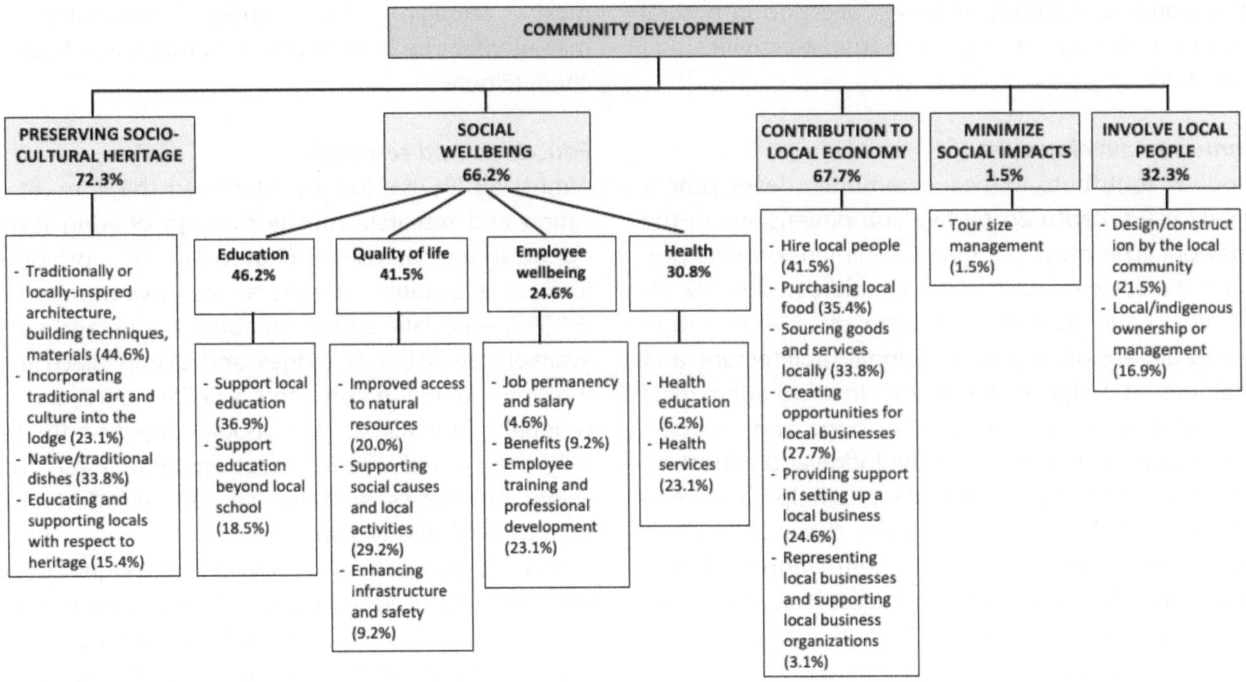

Figure 3. Local community development sub-dimensions and indicators and percentage of lodges reporting on them.

opportunities instead of unsustainable ones such as poaching, illegal mining, logging, or slash-and-burn farming. These new job opportunities can include guides, lodge staff, reserve rangers, para-biologists, handicraft artisans, or workshop leaders and can improve the quality of life for locals without engaging in activities that harm the environment. Value of conservation was also demonstrated regarding wildlife, where

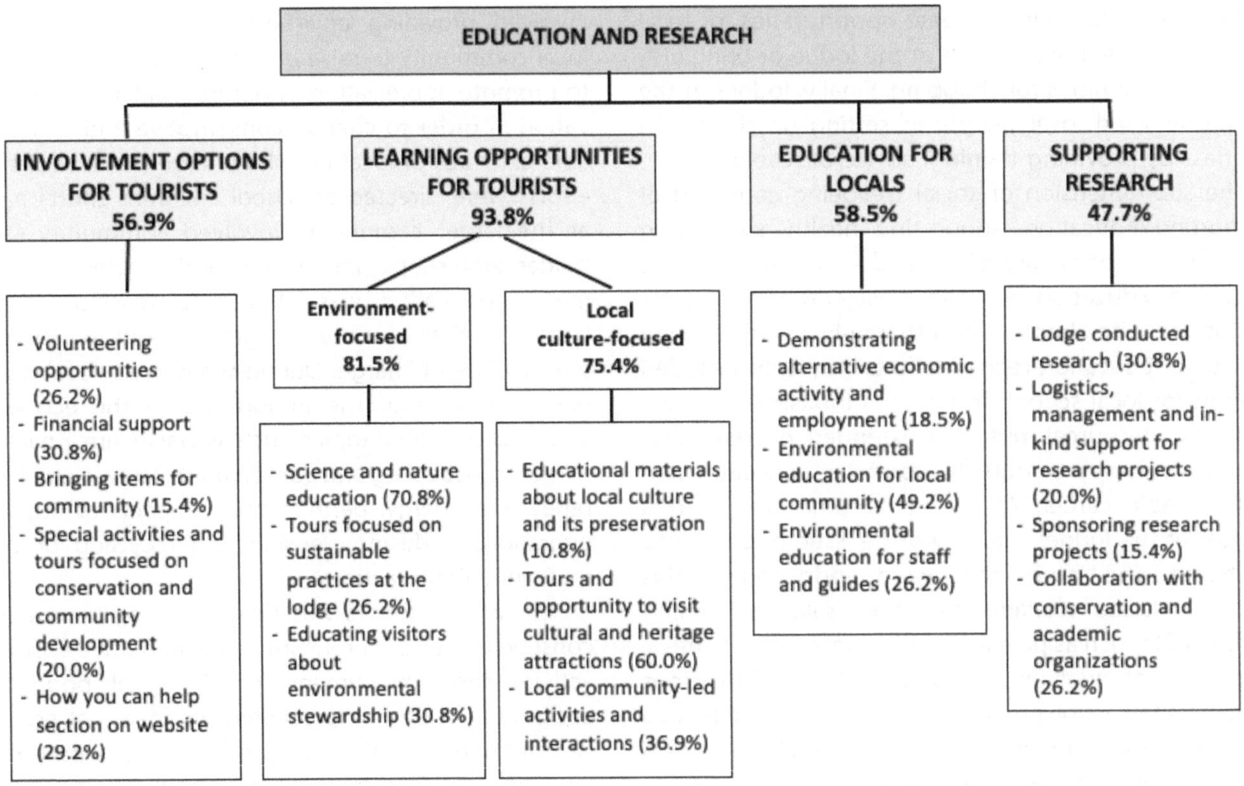

Figure 4. Education and research sub-dimensions and indicators and percentage of lodges reporting on them.

locals were shown that wildlife conservation contributes to the attractiveness of the area to tourists and can lead to benefits to the community. One of the lodges promoted the concept of financially viable conservation to local communities and landowners in order to encourage landowners to stop unsustainable farming practices and adopt a conservation and ecotourism model, thereby expanding the conservation area. The last theme in influencing local people's attitudes to conservation and ecotourism was environmental education for staff and guides, which focused mostly on environmental practices and responsible use of resources.

Opportunities for tourists to get involved in sustainability efforts at the lodge were available at 58.5% of the lodges. They included the option to donate money (30.8%), volunteer (26.2%) or participate in special activities and tours focused on conservation and community development, such as anti-poaching trips or data collection for research. One of the lodges promised a discount on fees to all visitors who volunteer in the community.

Finally, 47.7% lodges engaged in or supported research. The largest group of lodges (30.8%) conducted research themselves, for example, by monitoring or counting the wildlife in the reserve or by conducting socio-economic studies to help families living close to the lodge. Over a quarter of lodges (26.2%) collaborated with local and international conservation and academic organizations on research, while 20.0% provided logistics and in-kind support for research teams and 15.4% supported research projects financially.

Ecolodge profiles

Overall, the ecolodge assessment framework used in this study consisted of 5 dimensions and 79 indicators. The nature-based dimension was captured by three indicators, moral imperative by 5, environmental sustainability by 30, community development by 23, and research and education by 18 indicators. While all five dimensions were communicated by over 80% of ecolodges each, the indicators varied greatly in their communication level, see Table 3. Four categories of indicators were identified based on the communication level. All the ecolodges reported all three indicators from the 'nature-based' dimension. Other indicators with high communication levels included a commitment to sustainability and ecotourism, external recognition for sustainability efforts and availability of nature-based and education-based experiences sought by customers. On the other hand, ten indicators (4 from 'environment and conservation' dimensions and six from 'community development') dimensions were communicated by less than 10% of the ecolodges in the sample. These indicators included: choosing a location to minimize environmental impacts, reducing water pollution, reducing light pollution, garbage collection, tour size management, representing local businesses and supporting local business organizations, enhancing local community infrastructure and safety, providing job permanency and salary, providing employee benefits, and providing health education.

It should be noted that no indicators from environment and conservation dimension or community development dimension were communicated at a high level (by 60% or more ecolodges in the sample). The only indicators communicated at the high level were from 'nature-based', 'moral imperative' and 'research and education' dimensions, pointing to higher materiality of these issues from the perspective of ecolodges.

For each lodge, a total score in each category was calculated by counting the number of indicators present. The scores ranged from 5 indicators to 49 indicators. Figure 5 shows the number of ecolodges according to numbers of indicators addressed. The majority of lodges (56.9%) addressed between 11 and 30 indicators. Seven of the 65 operators communicated ten or fewer indicators, while only 7 of the ecolodges communicated over 40 indicators. None of the ecolodges communicated information on 51 or more indicators.

A cluster analysis was conducted to better understand different patterns of ecotourism communications in the sample. The results of the analysis indicated the presence of four distinct groups of lodges (Table 4).

Cluster 1 includes a group of top-performing ecolodges, which on average had the highest scores for communicating each dimension of ecotourism. They can be identified as '*engaged communicators*' with an extensive range of information provided for each aspect of ecotourism. In terms of organizational variables, all of these are individual properties, mostly located in desert or savanna biomes, with 60% of these in Africa. Cluster 2 includes '*selective communicators*', ecolodges that show high scores on communicating moral imperative, environment and conservation, and education and research dimensions, but lower levels of local community development communications. Most of these ecolodges are single properties located in forest biomes in South or Central America. Cluster 3 includes '*average communicators*', ecolodges that indicate a high level of moral imperative communication, and lower but similar scores on communicating environment and conservation, local community development, and education and research dimensions. These lodges are predominantly members of larger groups and are located in savanna and mountains biomes, mostly in Africa and Asia. Finally, cluster 4 includes a group of 18 lodges with the lowest scores

Table 3. The popularity of ecotourism indicators.

	Low communication Less than 20% of ecolodges	Medium-low communication 20% or more, less than 40%	Medium communication 40% or more, less than 60%	High communication 60% or more
Nature-based				• Location in natural settings • Accommodation and infrastructure with good access to nature • Nature-based activities
Moral imperative		• Contributions to social justice • Accountability (performance reporting) • Sharing profits with the community		• Commitment to sustainability/ecotourism • External recognition for sustainability efforts
Environment and Conservation	• Sewage treatment • Water harvesting and storage • Sustainable construction and energy-efficient design • Reducing environmental impact from tourist activities and operations • Buying sustainable food • Reducing noise pollution • Location choice to address environmental issues • Law enforcement and responders support concerning the environment • Garbage collection in natural areas • Reducing water pollution • Reducing light pollution	• Sorting, reusing and recycling waste • Reducing air pollution and carbon emissions • Energy conservation • Integrating the lodge with nature • Organic waste disposal • Water conservation • Reforestation, revegetation, weed control • Solar water heating • Collaboration with conservation organization and public sector on biodiversity conservation • Environmentally-friendly or sustainable materials • Financial contributions to conservation • Using sustainable supplies	• Use of renewable energy to power the lodge • On-site food growing • Restoration and preservation of wildlife • Sustainable land management and use of natural resources • Waste reduction • Using local materials • Wastewater treatment, reuse and water purification	
Community Development	• Support education beyond the local school • Local/indigenous ownership or management • Educating and supporting locals concerning heritage • Employee benefits • Enhancing infrastructure and safety • Health education • Job permanency and salary • Representing local businesses and supporting local business organizations • Tour size management to minimize social impacts	• Support local education • Purchasing local food • Sourcing goods and services locally • Native/traditional dishes • Supporting social causes and activities • Creating opportunities for local businesses • Support in setting up a business • Employee training and professional development • Health services • Incorporating traditional art and culture into the lodge • Design/construction by the local/indigenous community • Improved access to natural resources	• Traditional or locally-inspired architecture, building techniques, materials • Hire local people	
Research and Education	• Demonstrating alternative, ecotourism-based economic activity and employment to locals • Sponsoring research projects • Opportunity for visitors to bring items for community • Educational materials about local culture for visitors	• Community-led activities and interactions • Lodge-conducted research • Educating visitors about environmental stewardship • How you can help section on the website • Collaboration with other organization on research • Volunteering opportunities for visitors	• Environmental education for the local community	• Science and nature education for visitors • Tours and visits to cultural and heritage attractions

(*Continued*)

Table 3. Continued.

Low communication Less than 20% of ecolodges	Medium-low communication 20% or more, less than 40%	Medium communication 40% or more, less than 60%	High communication 60% or more
	• Environmental education for staff and guides • Logistics, management and in-kind support for research • Tours focused on sustainable practices at the lodge • Activities and tours focused on conservation and community • Opportunity for financial support from visitors		

on all the dimensions of ecotourism, except nature-based. This group can be classified as *'poor communicators'*. In terms of organizational variables, this cluster consists mostly of single properties that are located near water, such as private islands or lodges close to bays.

Discussion and conclusion

This study is the first to analyse, on a global scale, the Internet-based communication of ecotourism-related practices by leading ecolodge operators. It systematically examines the indicators of sustainability for both academic researchers and ecolodge operators and illustrates how different dimensions of sustainability are articulated in corporate communication via websites. The study confirms that ecolodge operators recognize the Internet as a useful conduit to communicate their activities and commitments, in agreement with observations made by Jones et al. (1998). The latter argued that the Internet allows small companies to reach an international audience. Although other studies have investigated ecolodges according to the use of defining characteristics of ecotourism (e.g. Lai & Shafer, 2005), these studies have been restricted by scale and focus. The current study included an international sample of ecolodges and avoided the inclusion of operators who self-declared themselves as an ecolodge, in favour of including ecolodges with independent third-party designations. The purposeful selection of ecolodges from Mehta (2010), World Travel Awards (2007–2019) and the National Geographic's Top 50 Ecolodges (Christ & Siber, 2008) was instrumental in this selection.

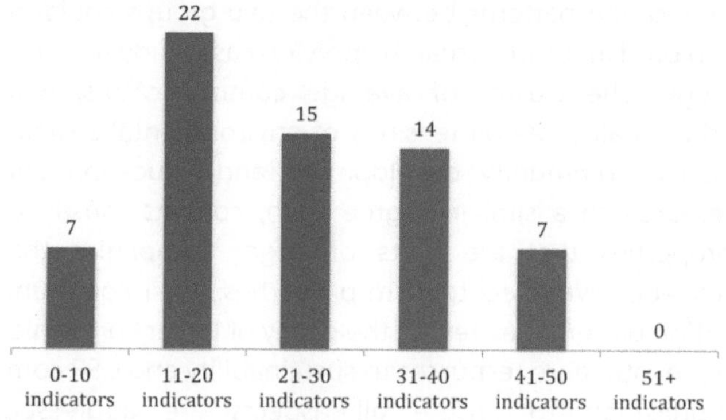

Figure 5. Number of ecolodges with frequency of indicators addressed.

The study found high levels of communication across all five dimensions of ecotourism discussed in the literature, i.e. a nature-based focus, the moral imperative, environmental sustainability, community benefits and development, and education and research (Fennell, 2020). This result was expected, given the approach taken to select the sample of leading ecolodges. Although the majority of ecolodges (46) provided some information about each ecotourism dimension, the study found significant variability in communication levels for different sustainability indicators within each dimension. While some indicators were communicated by 100% of the ecolodges in the sample, others were communicated by less than 5% of the sample. This wide range might point to differences in perceived materiality of related CSR issues, low adoption of some practices, a gap between theory and practice of sustainability and varying importance of different stakeholders. For example, efforts aimed at reducing light and water pollution were communicated by a very low number of ecolodges, possibly indicating that these issues are not recognized as a high priority by the sector and that ecolodges do not actively pursue practices aimed at meeting these goals.

Similarly, two employee wellbeing-related indicators had low communication level pointing to a lower priority for this issue. These findings indicate that some critical aspects of sustainability of ecolodges may have a low level of adoption, signaling a gap between theory and practice. A possible explanation can be proposed using stakeholder theory and the concept of stakeholder salience (Mitchell et al., 1997). In the present

Table 4. Cluster analysis: profiles of ecolodges.

	Cluster 1 (*n*=10) Engaged communicators	Cluster 2 (*n*=18) Selective communicators	Cluster 3 (*n*=19) Average communicators	Cluster 4 (*n*=18) Poor communicators
Nature-based	100%	100%	100%	100%
Moral Imperative	64.0%	45.6%	50.5%	27.8%
Environmental Sustainability	40.0%	30.2%	23.0%	21.9%
Community Development	38.3%	19.3%	23.8%	13.3%
Education and Research	45.0%	34.0%	26.9%	21.6%
Property type	100% single	88.9% single 11.1% group	94.7% group 5.3% single	88.9% single 11.1% group
Biomes	50% desert 30% savanna	94.4% forest	42.1% savanna 26.3% mountains	55.6% water-based
Location	60% Africa 20% Asia	61.1% South or Central America 16.7% Asia	47.4% Africa 26.3% Asia	38.9% Africa 22.2% Oceania or Caribbean

study, it can be noted that the majority of indicators with the highest level of communication address the informational needs of one group of stakeholders, the visitors. These indicators focus on the availability of features that ecotourists are interested in and include nature settings, environmentally friendly operations, biodiversity conservation efforts, availability of nature-based activities, environmentally based experiences and local culture-based education experiences.

On the other hand, several indicators with the lowest communication levels were aimed at employees and local community stakeholders. Jenkins (2004) pointed out that stakeholder relationships in small businesses are often based on a more informal, trusting basis and are characterized by personal engagement. Lähdesmäki et al. (2019) further conceptualized that the small business owner-managers' perceptions of stakeholders' saliency are influenced by the perceived social proximity between the stakeholders and the business. According to that theory, the local community and employees can be considered closer stakeholders than potential visitors, who often are geographically distant from the ecolodge. The findings of the current study indicate that ecolodges use the Internet mainly to communicate with the distant stakeholders such as potential visitors and less to communicate with the stakeholders who are close to them, such as employees. Indicators that are less mentioned may therefore reflect the limited interest in these issues on the part of both the ecolodges and the potential customers.

Cluster analysis of the sample indicated differences between ecolodges in terms of the focus and scope of their Internet communications. While there is a group of engaged communicators, providing more detailed information on different dimensions of ecotourism, other ecolodges are still lagging in the detail and extent of sustainability information they communicate on their websites. The presence of such ecolodges, with low scores on four ecotourism dimensions, possibly points to low adoption of ecotourism principles among this group. However, it may also indicate that several ecolodges do not fully take advantage of the Internet as a way to communicate their commitment to ecotourism principles, as was concluded in earlier research (Millar & Sammons, 2006). The indicators used in this study can serve as a guideline for ecolodges to either expand the range of initiatives they adopt or expand their communication also to include initiatives they may already have in place but fail to communicate using the Internet. The importance of Internet-based communication is exceptionally high for small operators who can use the Internet to reach potential customers and gain credibility.

Furthermore, the cluster analysis points to regional differences in how ecolodges articulate sustainability in their Internet-based communications. The group of selective communicators, who on average address environmental sustainability and education and research dimensions to a much higher degree than community development dimension includes mostly single properties located in forest biomes in South and Central America. Their relationships with local communities may display different characteristics than in case of the engaged communicators who predominantly are single properties located in desert and savanna biomes of Africa and Asia. Investigation of these relationships and the factors contributing to differences in communication patterns between the two groups could be explored in future research applying case study methodology. The cluster of average communicators, who address all three dimensions of environmental sustainability, community development and education and research to a similar degree each, consists mostly of properties that are parts of larger companies that manage several ecotourism properties. Their communication pattern may reflect the policy of larger companies to devote more resources to sustainability and CSR communication to ensure all aspects are addressed, although not to the same extent as engaged communicators address them.

The literature on CSR and CSR communication provides theoretical foundations and insight into motivating factors and benefits of CSR disclosure (Godfrey et al., 2009) which can be used better to explain the behaviour of ecolodges in the current study. We identified stakeholder theory (Freeman, 1984) as a core perspective in the relationship between the firm and stakeholders that are connected with the firm. Key to this theory is the need to engage stakeholders cooperatively and collaboratively in addressing the needs of multiple actors. We also used institutional legitimacy theory to explain the motivations for companies to disclose information on their sustainability practices to gain support from other stakeholders (e.g. community members and ecotourists). The literature contends that because of heightened levels of social and ecological concern on the part of the citizenry, those companies that communicate and disclose their CSR practices may gain more legitimacy and license for conducting their activities (Suchman, 1995). Our results should be interpreted in light of these two theoretical perspectives, as our research sheds light on many of the actual practices that are parallel to these theoretical perspectives. However, there is still considerable scope to engage in research that emphasizes more clearly the intrinsic versus extrinsic motivations that ecolodge operators embrace that ought to derivative from institutional legitimacy theory. This research should be conducted from inter-group and intra-group perspectives. By this, we mean that a firmer analysis of intrinsic versus extrinsic motivations for embracing CSR should be investigated within the ecolodge industry (intra-group dynamics) as well as between ecolodges and other forms of accommodation (inter-group). We would hypothesize, based on the findings reported here, that ecolodge operators ought to be more heavily invested in the intrinsic reasons for CSR practices than other forms of accommodation (see Fennell & Malloy, 1999), even though CSR may be used instrumentally for financial gain (Carroll, 1979; Lii & Lee, 2012).

Finally, our study has limitations. Most of the ecolodges in the sample were individually owned. However, in some cases, the ecolodges were part of a larger organization with several different locations. For some of them, it was not clearly stated what practices were implemented at which location. It is also essential to recognize that content analysis of communications published by companies focuses on self-reporting and represents a company perspective which not necessarily would agree with third-party interpretations of the company's practices or decisions.

Disclosure statement

No potential conflict of interest was reported by the author(s).

References

Alberici, A., & Querci, F. (2016). The quality of disclosures on environmental policy: The profile of financial intermediaries. *Corporate Social Responsibility and Environmental Management*, *23*(5), 283–296. https://doi.org/10.1002/csr.1375

Anderson, D. L. (1993). A window to the natural world: The design of ecotourism facilities. In K. Lindberg & B. Hawkins (Eds.), *Ecotourism: A guide for planners and managers, Volume 1* (pp. 116–133). The Ecotourism Society.

Baumann-Pauly, D., Wickert, C., Spence, J. L., & Scherer, A. G. (2013). Organizing corporate social responsibility in small and large firms: Size matters. *Journal of Business Ethics*, *115* (4), 693–705. https://doi.org/10.1007/s10551-013-1827-7

Björk, P. (2000). Ecotourism from a conceptual perspective, an extended definition of a unique tourism form. *International Journal of Tourism Research*, *2*(3), 189–202. https://doi.org/10.1002/(SICI)1522-1970(200005/06)2:3<189::AID-JTR195>3.0.CO;2-T

Bramer, M. (2020). *Principles of data mining*. Springer-Verlag.

Carley, K. (1993). Coding choices for textual analysis: A comparison of content analysis and map analysis. *Sociological Methodology*, *23*, 75–126. https://doi.org/10.2307/271007

Carroll, A. B. (1979). A three-dimensional conceptual model of corporate social performance. *Academy of Management Review*, *4*(4), 497–505. https://doi.org/10.5465/amr.1979.4498296

Ceballos-Lascuráin, H. (1996). *Tourism, ecotourism, and protected areas: The state of nature-based tourism around the world and guidelines for its development*. The World Conservation Union (IUCN), Gland, Switzerland; The International Ecotourism Society.

Christ, C., & Siber, K. (2008). 50 top ecolodges: The most earth-friendly retreats in the world's most spectacular wilds. *National Geographic Adventure*, *10*(9), 104–108.

Cromie, S., Stephenson, B., & Monteith, D. (1995). The management of family firms: An empirical investigation. *International Small Business Journal: Researching Entrepreneurship*, *13*(4), 11–34. https://doi.org/10.1177/0266242695134001

Dawkins, J. (2005). Corporate responsibility: The communication challenge. *Journal of Communication Management*, *9*(2), 108–119. https://doi.org/10.1108/13632540510621362

Deng, J., & Li, J. (2015). Self-identification of ecotourists. *Journal of Sustainable Tourism*, *23*(2), 255–279. https://doi.org/10.1080/09669582.2014.934374

DiMaggio, P. J., & Powell, W. W. (1983). The iron cage revisited: Institutional isomorphism and collective rationality in organizational fields. *American Sociological Review*, *48*(2), 147–160. https://doi.org/10.2307/2095101

Donaldson, T., & Preston, L. E. (1995). The stakeholder theory of the corporation: Concepts, evidence, and implications. *Academy of Management Review*, *20*(1), 65–91. https://doi.org/10.5465/amr.1995.9503271992

Donohoe, H. M., & Needham, R. D. (2008). Internet-based ecotourism marketing: Evaluating Canadian sensitivity to ecotourism tenets. *Journal of Ecotourism*, *7*(1), 15–43. https://doi.org/10.2167/joe185.0

Drumm, A., Moore, A., Soles, A., Patterson, C., & Terborgh, J. E. (2004). *Ecotourism development. A manual for conservation planners and managers. Volume II: The business of ecotourism development and management*. The Nature Conservancy.

Du, S., Bhattacharya, C., & Sen, S. (2010). Maximising business returns to corporate social responsibility (CSR): The role of CSR communication. *International Journal of Management Reviews*, *12*(1), 8–19. https://doi.org/10.1111/j.1468-2370.2009.00276.x

Elsbach, K. D., & Sutton, R. I. (1992). Acquiring organizational legitimacy through illegitimate actions: A marriage of institutional and impression management theories. *Academy of Management Journal*, *35*(4), 699–738. https://doi.org/10.5465/256313

Esrock, S. L., & Leichty, G. B. (1998). Social responsibility and corporate web pages: Self-presentation or agenda-setting? *Public Relations Review*, *24*(3), 305–319. https://doi.org/10.1016/S0363-8111(99)80142-8

Esrock, S., & Leichty, G. (2000). Organization of corporate web pages: Publics and functions. *Public Relations Review*, *26*(3), 327–344. https://doi.org/10.1016/S0363-8111(00)00051-5

Fennell, D. A. (2020). *Ecotourism* (5th ed.). Routledge.

Fennell, D. A., & De Grosbois, D. (2017). Sustainability initiatives of fly-fishing lodges: A global approach. *Acta Turistica*, *29*(2), 117–156. https://doi.org/10.22598/at/2017.29.2.117

Fennell, D. A., & Malloy, D. C. (1999). Measuring the ethical nature of tourism operators. *Annals of Tourism Research*, *26*(4), 928–943. https://doi.org/10.1016/S0160-7383(99)00032-8

Fitzgerald, M. A., Haynes, G. W., Schrank, H. L., & Danes, S. M. (2010). Socially responsible processes of small family business owners: Exploratory evidence from the national family business survey. *Journal of Small Business Management*, *48*(4), 524–551. https://doi.org/10.1111/j.1540-627X.2010.00307.x

Freeman, R. (1984). *Strategic management: A stakeholder approach*. Pitman.

Godfrey, P. C., Merrill, C. B., & Hansen, J. M. (2009). The relationship between corporate social responsibility and shareholder value: An empirical test of the risk management hypothesis. *Strategic Management Journal*, *30*(4), 425–445. https://doi.org/10.1002/smj.750

Gray, R. (2000). Current developments and trends in social and environmental auditing, reporting and attestation: A review and comment. *International Journal of Auditing*, *4*(3), 247–268. https://doi.org/10.1111/1099-1123.00316

Gray, R., Kouhy, R., & Lavers, S. (1995). Corporate social and environmental reporting: A review of the literature and a longitudinal study of UK disclosure. *Accounting, Auditing & Accountability Journal*, *8*(2), 47–77. https://doi.org/10.1108/09513579510146996

Gray, R., Owen, D., & Adams, C. (1996). *Accounting & accountability: Changes and challenges in corporate social and environmental reporting*. Prentice-Hall.

Grosbois, Danuta de, & Fennell, D. A. (2021). Sustainability and ecotourism principles adoption by leading ecolodges: learning from best practices. *Tourism Recreation Research*. https://doi.org/10.1080/02508281.2021.1875170

Hawkins, D. E., Epler Wood, M., & Bittman, S. (1995). *The ecolodge sourcebook for planners and developers*. The International Ecotourism Society.

Hendry, J. (2001). Economic contract versus social relationships as a foundation for normative stakeholder theory. *Business Ethics: A European Review*, *10*(3), 223–232. https://doi.org/10.1111/1467-8608.00236

Hill, L., & White, C. (2000). Public relations practitioners' perception of the world wide web as a communications tool. *Public Relations Review*, *26*(1), 31–51. https://doi.org/10.1016/S0363-8111(00)00029-1

Holder-Webb, L., Cohen, J. R., Nath, L., & Wood, D. (2009). The supply of corporate social responsibility disclosures among U.S. Firms. *Journal of Business Ethics*, *84*(4), 497–527. https://doi.org/10.1007/s10551-008-9721-4

Hou, J., & Reber, R. (2011). Dimensions of disclosures: Corporate social responsibility (CSR) reporting by media companies. *Public Relations Review*, *37*(2), 166–168. https://doi.org/10.1016/j.pubrev.2011.01.005

International Finance Corporation. (2004). *Ecolodges: Exploring opportunities for sustainable business*. IFC.

Isenmann, R. (2006). CSR online: Internet-based communication. In J. Jonkor & M. de Witte (Eds.), *Management models for CSR* (pp. 247–256). Springer.

Jenkins, H. (2004). A critique of conventional CSR theory: An SME perspective. *Journal of General Management*, *29*(4), 37–57. https://doi.org/10.1177/030630700402900403

Jenkins, H. (2006). Small business champions for corporate social responsibility. *Journal of Business Ethics*, *67*(3), 241–256. https://doi.org/10.1007/s10551-006-9182-6

Jones, K., Alabaster, T., & Walton, J. (1998). Virtual environments for environmental reporting. *Greener Management International*, *21*, 121–137.

Jones, T., Harrison, J., & Felps, W. (2018). How applying instrumental stakeholder theory can provide sustainable competitive advantage. *Academy of Management Review*, *43*(3), 371–391. https://doi.org/10.5465/amr.2016.0111

Jose, A., & Lee, S. M. (2007). Environmental reporting of global corporations: A content analysis based on website disclosures. *Journal of Business Ethics*, *72*(4), 307–321. https://doi.org/10.1007/s10551-006-9172-8

Kent, M., Taylor, M., & White, W. (2003). The relationship between Web site design and organizational responsiveness to stakeholders. *Public Relations Review*, *29*(1), 63–77. https://doi.org/10.1016/S0363-8111(02)00194-7

Kim, S. Y., & Park, H. (2011). Corporate social responsibility as an organizational attractiveness for prospective public relations practitioners. *Journal of Business Ethics*, *103*(4), 639–653. https://doi.org/10.1007/s10551-011-0886-x

Lai, P. H., & Shafer, S. (2005). Marketing ecotourism through the internet: An evaluation of selected ecolodges in Latin America and the Caribbean. *Journal of Ecotourism*, *4*(3), 143–160. https://doi.org/10.1080/JET.v4.i3.pg143

Lähdesmäki, L., Siltaoja, M., & Spence, L. J. (2019). Stakeholder salience for small businesses: A social proximity perspective. *Journal of Business Ethics*, *158*(2), 373–385. https://doi.org/10.1007/s10551-017-3707-z

Lewis, S. (2003). Reputation and corporate responsibility. *Journal of Communication Management*, *7*(4), 356–364. https://doi.org/10.1108/13632540310807494

Lii, Y. S., & Lee, M. (2012). Doing right leads to doing well: When the type of CSR and reputation interact to affect consumer evaluations of the firm. *Journal of Business Ethics*, *105*(1), 69–81. https://doi.org/10.1007/s10551-011-0948-0

Lyon, T. P., & Maxwell, J. W. (2011). Greenwash: Corporate environmental disclosure under threat of audit. *Journal of Economics & Management Strategy*, *20*(1), 3–41. https://doi.org/10.1111/j.1530-9134.2010.00282.x

Massi, M., & De Nisco, A. (2018). The internet-based marketing of ecotourism: Are ecotourists really getting what they want? In M. A. Camilleri (Ed.), *Tourism planning and destination marketing* (pp. 161–182). Emerald Publishing Limited.

McAdam, M., McAdam, R., Dunn, A., & McCall, C. (2014). Development of small and medium-sized enterprise horizontal innovation networks: UK agri-food sector study. *International Small Business Journal: Researching Entrepreneurship*, *32*(7), 830–853. https://doi.org/10.1177/0266242613476079

Mehta, H. (2007). Towards an internationally recognized eco lodge certification. In R. Black & A. Crabtree (Eds.), *Ecotourism series, number 5: Quality assurance and certification in ecotourism* (pp. 415–434). CABI Publishing.

Mehta, H. (2010). *Authentic ecolodges*. Harper Collins Publishers.

Mehta, H., Baez, A., & O'Loughlin, P. (2002). *International ecolodge guidelines*. TIES.

Meyer, J. W., & Rowan, B. (1977). Institutionalized organizations: Formal structure as myth and ceremony. *American Journal of Sociology*, *83*(2), 340–363. https://doi.org/10.1086/226550

Millar, M., & Sammons, G. (2006). A Content Analysis of Costa Rican Ecolodge Websites. *Hospitality Management*, Paper 11. http://repository.usfca.edu/hosp/11.

Mitchell, R. K., Agle, B. R., & Wood, D. J. (1997). Toward a theory of stakeholder identification and salience: Defining the principle of who and what really counts. *Academy of Management Review*, *22*(4), 853–886. https://doi.org/10.5465/amr.1997.9711022105

National Geographic. (2019). *Unique lodges of the world: sustainable tourism in action*. Sustainable tourism Impact Report.

Nekhili, M., Nagati, H., Chtioui, T., & Rebolledo, C. (2017). Corporate social responsibility disclosure and market value: Family versus nonfamily firms. *Journal of Business Research*, *77*, 41–52. https://doi.org/10.1016/j.jbusres.2017.04.001

Nelson, V. (2010). Promoting energy strategies on Eco-certified accommodation websites. *Journal of Ecotourism*, *9*(3), 187–200. https://doi.org/10.1080/14724040903144352

Nielsen, A. E., & Thomsen, C. (2009). Investigating CSR communication in SMEs: A case study among danish middle managers. *Business Ethics: A European Review*, *18*(1), 83–93. https://doi.org/10.1111/j.1467-8608.2009.01550.x

Oliver, C. (1991). Strategic responses to institutional processes. *Academy of Management Review*, *16*(1), 145–179. https://doi.org/10.5465/amr.1991.4279002

Osland, G. E., & Mackoy, R. (2004). Ecolodge performance goals and evaluations. *Journal of Ecotourism*, *3*(2), 109–128. https://doi.org/10.1080/14724040408668153

Pedersen, E. R. (2006). Making corporate social responsibility (CSR) operable: How companies translate stakeholder dialogue into practice. *Business and Society Review*, *111*(2), 137–163. https://doi.org/10.1111/j.1467-8594.2006.00265.x

Pollach, I. (2005). Corporate self-presentation on the WWW: Strategies for enhancing usability, credibility, and utility. *Corporate Communications: An International Journal*, *10*(4), 285–301. https://doi.org/10.1108/13563280510630098

Ross, W., & Wall, G. (1999). Ecotourism: Towards congruence between theory and practice. *Tourism Management*, *20*(1), 123–132. https://doi.org/10.1016/S0261-5177(98)00098-3

Russell, D., Bottrill, C., & Meredith, G. (1995). International ecolodge survey. In D. Hawkins (Ed.), *The ecolodge sourcebook for planners and managers* (pp. ix–xvii). The International Ecotourism Society.

Sangpikul, A. (2010). Marketing ecotourism through the internet: A case of ecotourism business in Thailand. *International Journal of Hospitality & Tourism Administration*, *11*(2), 107–137. https://doi.org/10.1080/15256481003732782

Scott, W. R. (2004). Institutional theory. In G. Ritzer (Ed.), *Encyclopedia of social theory, (pp.)* (pp. 408–414). Sage.

Sen, S., & Bhattacharya, C. B. (2001). Does doing good always lead to doing better? Consumer reactions to corporate social responsibility. *Journal of Marketing Research*, *38*(2), 225–243. https://doi.org/10.1509/jmkr.38.2.225.18838

Soundararajan, V., Jamali, D., & Spence, L. (2018). Small business social responsibility: A critical multilevel review, synthesis and research agenda. *International Journal of Management Reviews*, *20*(4), 934–956. https://doi.org/10.1111/ijmr.12171

Stemler, S. (2000). An overview of content analysis. *Practical Assessment, Research, and Evaluation*, *7*(17), 1–6. https://scholarworks.umass.edu/pare/vol7/iss1/17

Suchman, M. (1995). Managing legitimacy: Strategic and institutional approaches. *Academy of Management Review*, *20*(3), 571–610. https://doi.org/10.5465/amr.1995.9508080331

Sullivan, J. (1999). What are the functions of corporate home pages? *Journal of World Business*, *34*(2), 193–210. https://doi.org/10.1016/S1090-9516(99)00010-3

Vernon, J., Essex, S., Pinder, D., & Curry, K. (2003). The 'greening' of tourism micro businesses: Outcomes of focus group investigations in south East Cornwall. *Business Strategy and the Environment*, *12*(1), 49–69. https://doi.org/10.1002/bse.348

Wallace, G. N., & Pierce, S. M. (1996). An evaluation of ecotourism in amazonas, Brazil. *Annals of Tourism Research*, *23*(4), 843–873. https://doi.org/10.1016/0160-7383(96)00009-6

White, C., & Raman, N. (1999). The worldwide web as a public relations medium: The use of research, planning and evaluation in web site development. *Public Relations Review*, *25*(4), 405–419. https://doi.org/10.1016/S0363-8111(99)00027-2

Wight, P. (1997). Ecotourism accommodation spectrum: Does supply match the demand? *Tourism Management*, *18*(4), 209–220. https://doi.org/10.1016/S0261-5177(97)00008-3

Sustainability for ecotourism: work identity and role of community capacity building

Clare D'Souza, Mehdi Taghian, Timothy Marjoribanks, Gillian Sullivan Mort, Md Manirujjaman and Stephen Singaraju

ABSTRACT

This research sheds light on community organisations and how they can shape the landscape for ecotourism sustainability. It identifies work identity attributes that involve pro-environmental behaviours for achieving institutional challenges to meet sustainability for ecotourism goals. First, a framework on managing community organisations and its impact is proposed and empirically tested, which includes theoretical perspectives of work identity attributes like pro-environmental behaviours, community values and involvement for building community capacity. A conceptual model was developed to guide this study. A mixed-method study has been designed involving 28 interviews, followed by testing 266 responses, analysed through Structural Equation Modelling and Cluster Analysis. Through identifying clusters of individuals based on their demographics, behaviours and perceived individual effectiveness for supporting ecotourism the research demonstrates that individuals with pro-environmental behaviour can act as agents of change for realising ecotourism and sustainable development goals. The research highlights that work identities require specific personal attributes to influence the change process through sustainability for ecotourism. By examining these attributes on community values and involvement, this study offers insights for institutional change processes that meet the central tenets of sustainability for ecotourism.

Introduction

According to Blamey (2001), ecotourism has three dimensions, nature-based, environmentally educated and sustainably managed and observes two sustainability principles (i) to support local economies and (ii) to support conservation. Thus, ecotourism and sustainable tourism can be analogous, in a sense that ecotourism encapsulates sustainable tourism. Ecotourism is a subcomponent in the domain of sustainable tourism and aims to achieve sustainable development by planning and managing successfully to achieve its social and environmental objectives (Wood, 2002). For example, King and Stewart (1996, p. 293) have contended that the expectations of ecotourism 'is the formation of a symbiotic relationship between tourist, indigenous people and natural areas'. They propose that sustainable relationships in connection to the environment and developments of local welfare should transpire (Winson, 2006). Although, sustainable tourism and ecotourism are not synonyms, as there are many avenues of ecotourism which may not be sustainable, and if sustainability for ecotourism is to be upheld, then cautious planning and management is essential (Wall, 1997).

Sustainability for ecotourism is universally and equally valuable to society but at the same time it is debated within existing research on whether the concepts are of theoretical significance only, or whether they are academic reflections as contrasted to empirically supported realities (Simpson, 2001). In terms of community involvement, community adoption rates for the sustainability of ecotourism has been slow, yet, there is some merit in this concept to warrant continuing efforts concerning its effective realisation (Simpson, 2001). While several scholars have made notable contributions to community participation for tourism development (Murphy, 1985; Reed, 1997; Simmons, 1994), relationships between work identity and community organisations remain largely unexplored. Developing a community work identity institutes, the context for assessing potential influences and are critical for project decision-making.

For efficiency and effectiveness involving sustainability for ecotourism, amongst other factors also requires the support and participation of residents of a community (Fallon & Kriwoken, 2003; Nicholas, Thapa, & Ko, 2009). Tourism development is not achievable without the involvement of communities (Richards & Hall, 2003). The view that community organisations may be an important source driving ecotourism has its roots in the residential structure of the community. Work identity involving residents who are pro-environment, having a strong sense of environmental sustainability, are the champions of this industry who may be in a better position to contribute to this transformation. Against this backdrop, the research identifies work identity attributes that involve pro-environmental behaviours to form an important composite in achieving the complexities of institutional changes for meeting sustainable ecotourism goals. Hence, this research draws attention to the community organisations for changes in this translation process by proposing a framework and uses a mixed-method study to seek empirical evidence of, and to estimate the relationships hypothesised.

The aim of the qualitative interviews is to explore and clarify issues surrounding the community's level of support for ecotourism sustainability and examining work identities to better manage sustainability activities. Following which, the research was designed to generate a theoretical framework to understand and identify relationships between work identities and community organisations. Understanding work identity influences on community interactions for achieving sustainability goals within the context of ecotourism will provide broader perspectives for managing community organisations; secondly, it aims to identify specific market segments and finally, by examining these attributes on community values and involvement, this study offers insights for institutional change processes that meet the central and overarching tenets of sustainability.

This research makes the following contributions. First, a framework is proposed and empirically tested, which includes theoretical perspectives on work identities. Second, the research identifies potential clusters of individuals based on their demographics and perceived individual effectiveness that are likely to support sustainability for ecotourism. Third, this research provides a better theoretical understanding on the role of work identities that involve attributes such as pro-environmental behaviours like nature-related behaviour and sustainable tourist behaviour. By engaging with these issues, this paper contributes to the conceptual and empirical understanding of community organisations' capacity building within the context of ecotourism sustainability.

Theoretical framework

United Nations World Tourism Organization is working with governments and other stakeholders to achieve the revised 2030 Agenda for Sustainable Goals, of which several of them place an emphasis where tourism is featured. The subsequent thrust is about building community capacity. Engaging with questions of ecotourism for sustainability successfully requires local and inclusive participation, buy-in, and action. Communities can perform as change agents to obtain social and behavioural outcomes (Butterfoss, 2006). Keys, Thomsen, and Smith (2016), in their social change research, found that specific individuals perform substantial roles in communicating innovative ideas and practices; this has been illustrated in community health, sustainable practices and sustainable technologies. These specific individuals in a community can be community residents acting as change agents or community leaders. The inclusion of specific community members is fundamental for planning and managing ecotourism activities, and failure to understand community participation, or exclusion of such participation, can hinder programme outcomes (Butterfoss, 2006). Thus, identifying the type of community residents who should engage with ecotourism community organisations will help to define work identities that make a difference to the outcome of sustainability for ecotourism capacity building programmes.

Can tourism be successful through receiving residents' support? On the other hand, opponents disagree with community perspectives for local control of the industry (Blank, 1989). While some do not reference local involvement within a global setting (McIntyre, Hetherington, & Inskeep, 1993). Joppe's (1996, p. 475) perspective on community participation suggests there has to be broad participation by all members of the community, and defines community as 'self-defining, in that it is based on a sense of shared purpose and common goals. It may be geographical in nature or a community of interest, built on heritage and cultural values shared among community members'. In contrast, Taylor (1995) suggests that community objectives may not be possible. While community consultation may be present in many cases, community participation is sporadic. Even if community participation was involved, practical difficulties have resulted in inadequate outcomes leading to poor quality decision-making (Simpson, 2001). This research argues that for a better-quality decision making there is essentially a need for having strong work identity attributes, which relates to pro-environmental community action, requiring more knowledgeable residents to influence situations relating to sustainability for ecotourism.

Simmons (1994) suggests that there are additional challenges in planning and implementing community participation in tourism planning. For example, first, that the process of tourism development requires a high degree of early involvement and acceptance from residents. While this process takes time, without satisfactory resident involvement, uncertainties and misunderstandings are likely to evolve. Second, finding the right type of community members to represent the community's commonly held views to engage in tourism planning can be problematic. Local residents differ in their perception and those with pro-environmental perspectives are likely to shape planning of ecotourism, meaning other voices or perspectives may be excluded. Thus, a scoping study was undertaken to understand the issues raised within sustainability for ecotourism and community engagement at the local community level, following which a framework is presented.

Qualitative study

The qualitative phase of the study involved in-depth discussions with a group of active community residents regarding sustainability for ecotourism, climate change and building community capacity. These interviews were held in two states of Australia, Queensland (7 respondents) and Victoria (15 respondents). The discussions went from the more general questions regarding climate change and responsible tourism to more specific issues, such as community resilience, community Involvement and engagement and residential interactions to determine work identities for communities. Protection of communities and strengthening resilience of the economy was discussed, as well as the need for adaptation to manage risks from climate change impacts by governments at all levels including, individuals, groups, the role of institutions and business organisations. Informed by some responses of the discussions and the literature review, a framework was developed.

In addition, six interviews were held with city councils' planning executives to determine their views on community engagement for ecotourism development within community organisations. Since we intended to test a theory with directional hypotheses, a standardised structured interview method was used (Fowler & Mangione, 1990). The interviews provided an understanding of community relationships and its involvement in ecotourism. Four of the six councils claimed that they have already kept ecotourism high on their agenda as they had a high rate of tourist traffic in their localities.

The analysis was inductive, informed by the literature. It allowed categorising items to create themes (Miles & Huberman, 1994). Summary of the scoping study resulting from the qualitative interviews for involving communities were as follows:

- Degradation of the local areas: the respondents believed that degradation of the local areas, more particularly heritage-listed structures would lead to loss for the community rather than loss for individual residents. Community Involvement came out more strongly.
- Lack of knowledge of ecotourism: the respondents expressed that many community members are not aware of ecotourism related issues, such as its longer-term impact on the 'natural wealth of the community' and 'sustainable tourist activities'. Some respondents suggested this was the 'biggest drawback on the tourism system'.
- Understanding of 'ecotourism': they expressed that many community members were not well informed of the outcomes of 'ecotourism', and its relation in terms of degradation of local ecological, economic and social systems. On the other hand, those that were aware associated the concept to nature and wellbeing to sustainability. While others associated ecological footprint.
- Educating residents: respondents believed that well-informed residents with the 'right attitude' and 'skill set' could emerge as the main driver for ecotourism. The responses from the council interviews were also supportive of this theme and had made several attempts for public education to get the communities interested in participation.
- Fostering pro-environment sentiments: the community and council interviews responses for ecotourism participants was 'pro-environmental behavior and views to get moving'. This should be high on the agenda with 'participants involved in communities', followed by 'engagement', 'interest in sustainable issues' 'support for ecotourism' 'connection with the environment' 'making structural changes to adapt to climate change' 'capacity building' 'self-determination' and 'ecological awareness'. Strongly motivated and pro-environmental individuals were the central force for driving sustainability. Many of these abilities were lacking within community participants.

An appraisal of the community's tourism issues and decisions with respect to public participation raised questions about who the environmentally concerned residents and their attitudes towards ecotourism would be? (Haywood, 1988). Even though large community projects recognise several factors for successful community participation, Butterfoss (2006) has found that the type of individuals or representative groups that will serve as community participants for ecotourism development is well defined. This research has found that tourism development in this area has been slow to recognise and profile work identities for those individuals and groups. While these population segments can be well informed in sustainability matters by industry standards, yet, if excluded from the development process may result in potential risk of not being able to develop sustainability for ecotourism agreements and overcoming possible resistance from various stakeholders. This could jeopardise the attempt to maximise the sustainability efficiency and benefits expected. Specifically, the objective of the interviews was to clarify and identify issues surrounding the community's level of support for ecotourism sustainability and defining work identities to better manage sustainability activities.

Identities are believed to be the qualities, beliefs and personality that make a person (James, 2015). Walsh and Gordon (2008, p. 46) has referred to individual work identity as 'a work-based self-concept, comprised of a combination of organizational, occupational, and other identities, that affect the roles people adopt and the corresponding ways they behave when performing their work'. Work identity is somewhat cognitive as it describes the qualities and conception of what individuals hold of themselves at work (Wrzesniewski & Dutton, 2001).

Identities are vital to understanding institutional change. It is not surprising to find how identity work can alter dialogues, organisations and societies (Giorgi & Palmisano, 2017; Leung, Zietsma, & Peredo, 2014; Wright, Nyberg, & Grant, 2012). Brown's (2018) view on identity in local settings has been derived form a collection of strategies to establish context definite characteristics of identities (Gabriel, Gray, & Goregaokar, 2010). For instance, Wright et al. (2012) have classified the identity positions of sustainability managers into 'the green change agent', the 'rational manager' and the 'committed activist'. Wright's et al. (2012) examination on Australian sustainable mangers identified how *identity* work operated to expand the knowledge about climate change and influence on others with the possibility of inspiring and changing discourses. Community organisations are capable of mobilising these discourses by prioritising on having the right type of identity to broadly circumspect political and social change.

Thus, our research question – should work identity define profiles of individual characteristics that have enough latitude to exercise a better delivery of a coordinated sustainability programme for action?

The following are the key findings for defining work identity attributes involving pro-environmental behaviour:

(1) To understand work identities the following question was asked: Why is it necessary to involve pro-environmental residents in community participation/capacity-building for supporting ecotourism development? Responses are categorised into two groups – external and internal, with some key responses:

External relevance

- Increasing social pressure and trends globally to protect the environment
- No adequate regulations to support ecotourism development
- Strong need to raise awareness within communities
- Avoidance of bureaucratic decisions and revert to proactive hands on decisions

Internal relevance

- Pro-environmental residents possess qualities of a good corporate citizen
- Attachment to nature
- Strong commitment to betterment of the environment
- Individual's enthusiasm to do good for the environment
- Individual's willingness to be involved for a common good cause

(2) How can local councils draw communities to implement initiatives to support ecotourism through work identity attributes that involve pro-environmental characteristics? Responses involve:

Capacity development

- Assist participants with engaging individuals in social, behavioural change which leads to sustainable development
- Allow participants to strengthen the skills, competencies that involve communities to foster sustainable growth
- Permit interactive participation

Council and community engagement

- Councils should fund and effectively allocate revenues
- Implement and support initiatives to minimise tourist ecological footprint

- Inspire communities and pro-environmental individuals by leading through example
- Aim to create a network of stakeholder partnerships and build national communities
- Promote and exercise sustainable practices wherever possible

(3) What reasonable outcomes would you expect with work identities that involve pro-environmental resident's support for sustainable development? Responses involve:

- Pro-environmental residents are likely to facilitate better performance and efficiency towards the progress of sustainable development
- They are likely to be more knowledgeable in terms of environmental impact
- Higher level of accountability
- Possibility of influencing others due to their positive attitudes towards the environment

To better understand how communities can transform substantial sustainability-level changes to managing and delivering ecotourism objectives we propose the following framework based on the above findings and literature.

Discussion qualitative analysis

The above framework in Figure 1 shows that community resilience and community rigidity would depend on the intensity of community interaction and their effects. Drawing from the work of Heinze, Soderstrom, and Heinze (2016), communities are considered to be more local, having a high concentration of social and economic interdependence with local actors in a communal geographic setting (Fligstein & McAdam, 2012; Marquis, Glynn, & Davis, 2007). They are influenced by several systems at the macro level that is external. For instance, the social normative system that underpins social networks set social and cultural standards and conformity (Galaskiewicz, 1997; Marquis, 2003). The political system enforces regulation and political mandates (Guthrie & McQuarrie, 2005). Community systems are local actors that have community interactions sharing knowledge, control, upkeep and governance on how and what is required within this system.

At the micro-levels i.e. internal, community actors, in this case it is the individual that brings about changes through their own identities, commitment and interests (Lok, 2010). In parallel, social interaction through organisational members create shared meanings and develop practices to adapt institutional change to local environment, sustaining both institutional restrains and local interests (Hallett, 2010; Heinze et al., 2016; Kodeih & Greenwood, 2014).

The qualitative interviews found that communities were not well equipped to deal with the issues surrounding sustainability for ecotourism. They believed that all stakeholders should have shared interest in sustainability for ecotourism, but local community organisations can make the greatest contribution, which we identify would be through their interactions. Not all respondents had faith in ecotourism but suggested it requires expertise from within the community. This was an interesting comment, as it indicated that these respondents believed in relying on communities to further initiate and innovate actions towards ecotourism.

When asked to define work identity that refers to 'pro-environmental resident' in the context of what type of residents they believed could take initiatives and drive future support for ecotourism, many indicated attributes such as 'tourism' related sustainable behaviour, while some added 'emotional attachment' or the 'feel good do good' to the environment.

This qualitative study is exploratory and formative in nature. The qualitative research showed there is a lack of community capacity building, which is the process of creating, and growing the skills, competencies and processes that community organisations require to endure, adjust and prosper to meet sustainable goals.

Quantitative analysis

The above findings explain why it is necessary to consider work identity attributes that involve pro-environmental individual behaviours for fostering sustainability for ecotourism, supported by community involvement and community values. Based on findings from the present qualitative study, we investigate the relationship between work identities involving attributes such as pro-environmental and sustainability-led behaviour to foster ecotourism community growth.

The qualitative research was followed by a quantitative study. Two analytic methods are used in the study, Structural Equation Modelling and Cluster Analysis. The research instrument was developed using the scales adopted from the literature and verified by some of the findings of the qualitative study. The questionnaire was pre-tested before its administration. A professional market research company was commissioned to conduct the pre-test of the questionnaire and the data collection. The pilot test consisted of 20 respondents to pre-test the instrument (Baker, 1994) for clarity of the questions and the sequence of questions (De Vaus, 1993). Minor adjustments to the wordings of some of the questions were deemed necessary.

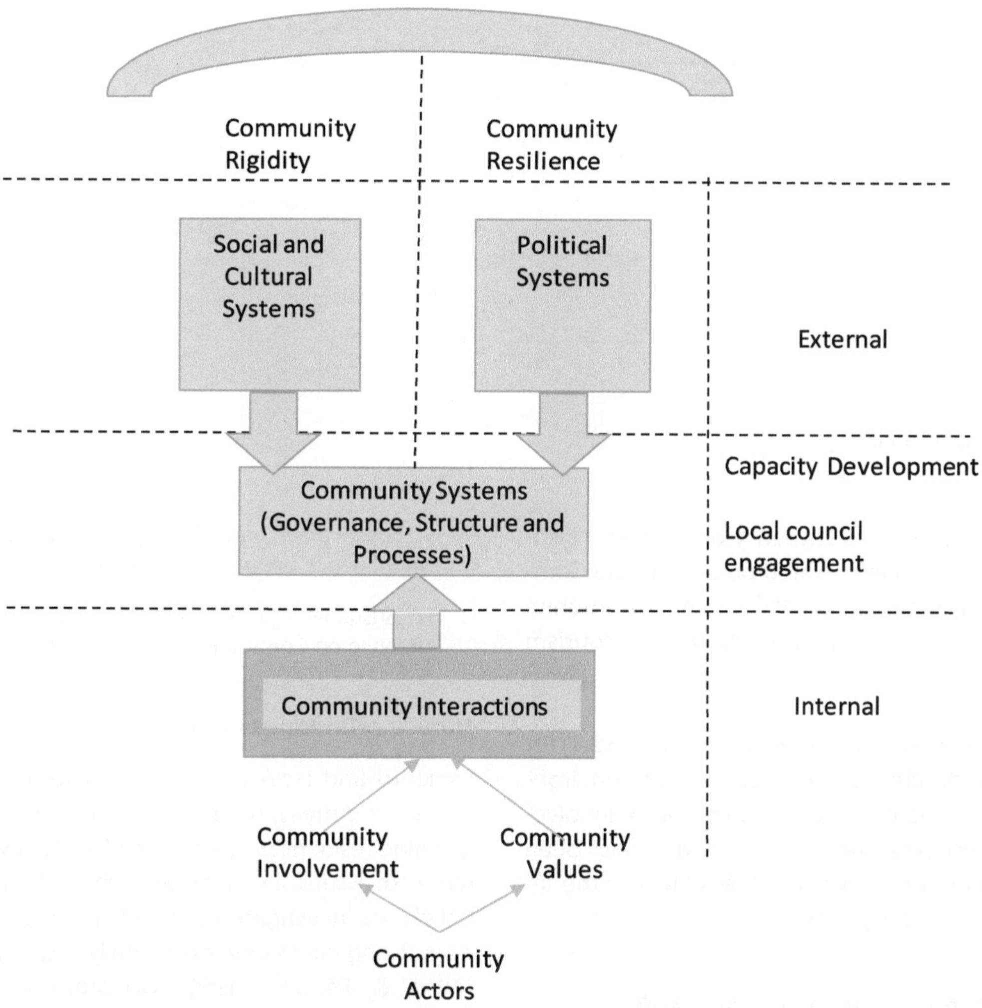

Figure 1. Community framework sustainability for ecotourism: managing and impact.

This study uses Australia as the country of analysis. The data for this study were collected from a national Australian online survey of 266 respondents sourced through a commercial online panel operator. Respondents were located across all Australian states and territories, roughly in proportion to their share of the national population.

The sample consisted of 46% males and 54% females. The age groups were: 18–24 (11%); 25–34 (19%); 35–44 (16%); 45–54 (17%); 55–64 (17%) and 65+ (20%). The study was conducted in Australia, with most of the respondents coming from NSW (32%), followed by Victoria (27%); Queensland (20%); South Australia (7%); Western Australia (9%); ACT (4%); Tasmania (6%) and Northern Territory (2%), very close to their share of national population. In terms of education, most of the respondents had either a TAFE/Technical certificate or Diploma (33%) or completed high school (31%). There were fewer post-graduate degrees or higher respondents (11%) than university (undergraduate) degree (23%). A few respondents fell in the other category (2%). Many of the respondents were employed full time (30%); followed by part-time (18%). A small number of respondents reported that they were self-employed (5%) and a few of them were also unemployed (5%). Retirees comprised of 23% of the sample, students 5%; home-maker 12%, and the other category represented 2%.

Structural equation modelling

In understanding the current sense of community, we employ two types of behaviour of those residents, one that displays interest toward 'sustainability in tourism' behaviour and other 'nature' related behaviours. There is a gap in tourism literature regarding sustainable behaviours influence through communities and support of sustainability for ecotourism, thus raising the following key research questions:

- To what extent does work identity attributes such as sustainable related tourist behaviour and nature relatedness influence the level of Community Value and Community Involvement?

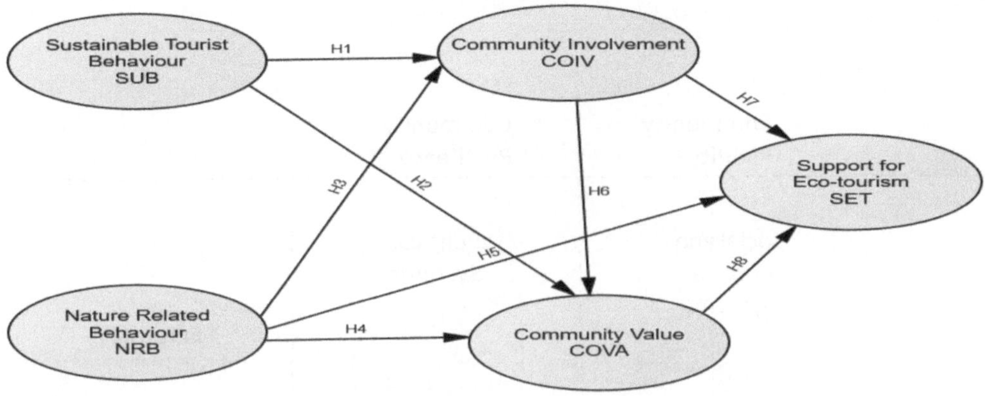

Figure 2. Conceptual model.

- Is there an association between work identity attributes such as sustainable related behaviour and Community Involvement and Value in community relationships that help support future ecotourism activities?

Figure 2 demonstrates the conceptual model for examining work identity attributes such as pro-environmental behaviour relationships on Community Involvement and Community Value. The model has been designed based on interviews and the review of the literature presented in this paper.

Sustainable related tourist behaviour- SUB

Having a pro-environmental attitude does not necessarily mean making environmentally ecotourism choices (Juvan & Dolnicar, 2014). Researchers have found, for example, that people do not change their vacation behaviour due to climate change (Hares, Dickinson, & Wilkes, 2010). Others have found that, even if they are aware of climate change, some individuals are unlikely to change behaviour (McKercher, Prideaux, Cheung, & Law, 2010). Nevertheless, some research found that consumers are making sound decisions on the sustainability of everyday products and are also eager to transfer those habits to purchasing tourism products (Miller, 2003), while environmentally aware consumers may be more likely to have pro-environmental behavioural intentions than others (Lee & Moscardo, 2005). The low support from consumers is suggested to be a hindrance for the growth of ecotourism (Budeanu, 2007). Behaviours for example that a pro-environmental resident would seek are effective mobility schemes, reducing half the travel distance or the use of eco-efficient accommodation (Budeanu, 2007; Martens & Spaargaren, 2005). Thus the relationships between sustainably sound behaviours and community involvement and values is necessary. This leads us to the following hypotheses:

H1: Sustainable Tourist Behavior (SUB) has a direct positive influence on Community Involvement (COIV)

H2: Sustainable Tourist Behavior (SUB) has a positive influence on Community Value(COVA)

Nature relatedness – NRB

Lankford and Howard (1994) indicate that a number of studies and measurement procedures have been used to examine resident perception and attitudes in relation to a range of issues. Even motivation, attitudes, values, and beliefs are investigated to identify why people engage or do not engage in environmentally responsible behaviour (Allen & Ferrand, 1999; Nordlund & Garvill, 2002). However, research shows nature relatedness is an influential force for environmental protection (Bragg, 1996; Conn, 1998; Feral, 1998; Saunders, 2003). It has increasingly become beneficial in understanding pro-environmental behaviours (Zelenski & Nisbet, 2014). Nature relatedness is akin to other environmental measures in predicting pro-environmental attitudes and behaviours. Nature relatedness scales assess the affective, cognitive and experiential characteristics of individuals and their connection to nature (Nisbet, Zelenski, & Murphy, 2009). Mayer and Frantz's (2004) study on connectedness to nature demonstrated that ecologists (Leopold, 1949; Pretty, 2002) and eco-psychologists (Roszak, 2001; Fisher, 2002) have identified that connection to nature is the main driver of encouraging ecological behaviour. Similarly, as relationship closeness to nature increases, empathy and willingness to help also increases (Mayer & Frantz, 2004).

Disconnecting from nature may result in environmental destruction, as individuals may not be encouraged to protect the environment (Nisbet & Zelenski, 2011; Schultz, 2000), while on the other hand a closer relationship with nature is contingent on experiences in natural environments (Nisbet & Zelenski, 2011; Ward Thompson, Aspinall, & Montarzino, 2008). Thus, a high level of nature relatedness or a strong subjective association with nature is positively associated with greater environmental concern and thus potentially has important support implications (Nisbet & Zelenski, 2013). Nisbet et al. (2009) found that nature relatedness captures specifically three areas of human-nature relationships – cognition, affect, and experience and captures individual differences far better than other differences in environmental relationship behaviour. It seems clear that the key principles of near relatedness to nature will be more important to community engagement in ecotourism than attitudes and opinions. Thus, we propose the following hypotheses:

H3: Nature Related Behavior directly and positively influences Community Involvement(COIV)

H4: Nature Related Behavior directly and positively influences Community Value (COVA)

H5: Nature Related Behavior has a direct positive association with future support for Ecotourism (SET)

Community involvement – COIV

Community participation in tourism planning can involve various stakeholders, including local government officials, local citizens, architects, developers, business people, and planners (Haywood, 1988). Taylor (1995, p. 487) has supported the concept of Community Involvement for developing ecotourism and argues, 'the concept of community involvement in tourism development has moved nearer to the center of the sustainability debate'.

Community tourism planning for sustainability involves implementation and control. Some of the major components of ecotourism are local community participation and environmental sustainability that are associated to support tourism (Choi & Murray, 2010). Not only does it analyses the economic, ecological and informational components of strategy but links itself to the social and political factors inside and outside of the industry (Haywood, 1988). As with any planning, sustainability is meaningless unless it is recognised and executed at the operational level. Thus, residents need to be more acceptable by recognising sustainability to enable them to transfer this power for exerting influence.

In order to enhance sustainability development in tourism, it is necessary to involve the local community from the onset of the development stages (Murphy, 1981). It is argued that multiple stakeholder participation would be an effective model in a strategic oriented planning process (Simpson, 2001). Thus, Community Involvement in essential ecotourism led activities may support future ecotourism suggesting the following hypotheses:

H6: Community Involvement has a direct and positive influence with Community Value

H7: Community Involvement has a direct and positive influence with future support for Ecotourism

Community value – COVA

Notably, any development process initiated through tourism should not undermine the values residents hold within the community (McCool & Martin, 1994). Pro-environmental residents will express their value of respect towards nature by participating in activities that protect the environment (Fraj & Martinez, 2006). That is, individuals who see value in ecological issues have a higher propensity to participate in environmental behaviour (Alonso, 1999). An examination of value-based environmental beliefs and behaviour showed that people are generally likely to engage in pro-environmental behaviours (Nordlund & Garvill, 2002; Schultz & Zelezny, 1998). Furthermore, environmental concern measured by the New Environmental Paradigm scale also shows its association with pro-environmental behaviour (Dunlap & Van Liere, 1978; Dunlap, Van liere, Mertig, & Jones, 2000).

Values is a key element in explaining behaviour. In management they affect the management structure and strategies for success (Bamberger, 1983). There are several classifications proposed for values. For instance, Rokeach's (1973) Value system for terminal and instrumental values, while Homer and Kahle (1988) reclassified Rokeach's values in those internally oriented. Schwartz (1992), on the other hand, emphasises the social level in the field of intercultural research. Personal norms are related to an individual's value system and have a greater temporal stability than an average intention to behave (Klockner & Blobaum, 2010).

Durkheim (1964) posited different types of community relationships, one founded on shared interest and values and the other founded on functional interests and impersonalisation. While Burroughs and Rindfleisch (2002) recognised two perspectives from the literature (Berlin, 1997; Davidson & Cotter, 1997; Glynn, 1986), they propose that community can be viewed as a geographically bounded space. Based on this view of community, we define Community Values as the value placed on developing and maintaining relational ties with one's local geographic community (Burroughs & Rindfleisch, 2002, p. 357) and thus use the Community Value Scale to measure Community Value. This leads us to the following hypothesis:

H8: Community Value has a direct and positive relationship with future support for Ecotourism

Measurement

Scales used in this study were drawn from several sources found in the literature. Measurement scales comprise of multi-item five-point Likert-type scales anchored by 'strongly agree' and 'strongly disagree', these statements are illustrated in Table 1. The two constructs for pro-environmental behaviour are listed as Sustainable Tourist Behaviour and Nature Readiness Behaviour. The items for Sustainable Tourist Behaviour were taken from McKercher et al. (2010), and for Nature Readiness Behaviour were adopted from Nisbet et al. (2009). These items enabled predicting specific types of environmental activism and identity better than many of the other existing scales (Nisbet et al., 2009). Relatedness scale is a self-reported measure designed to evaluate the affective, cognitive, and physical relationships of individuals in relation to nature. It has been found to be a better predictor of involvement in environmental groups, sustainable consumption and identification as an environmentalist (Nisbet et al., 2009).

To measure Community Involvement and Community Value, scale items were adopted from Lee (2013) and Burroughs and Rindfleisch (2002) respectively. There is a dearth of available measures for Community Values, and thus, since Burroughs and Rindfleisch (2002, p. 357) developed a new measure, these measures were utilised. Their measures as guidelines hinged on three key components from Davidson and Cotter (1997), i.e. (1) a belief in providing input and ideas on important issues facing one's community, (2) a willingness to contribute time and effort to community improvement, and (3) a willingness to contribute financially to the betterment of one's community. Following which they developed a new nine-item Community Values scales that assesses the above three points'. The willingness to support ecotourism was modified to suit this research and was taken from Thøgersen, Haugaard, and Olesen (2010).

We conducted an exploratory factor analysis tests for data reduction and a consistency check on the survey questions to develop the variables used in the model (Figure 1). Twenty-one items were retained after conducting the confirmatory factor analysis using AMOS, dropping some to obtain a good model fit. We performed an Inter-item and item to total reliability tests (Churchill, 1979) as depicted in Table 1.

As per Table 1, Cronbach alpha ranges from 0.71 to 0.93 for the five variables. The average variance extracted (AVE) ranged from 0.52 to 0.82 and Composite reliabilities (CR) range from 0.77 to 0.93. The effectiveness of alpha values where 0.7 is good and 0.80 plus is respectable DeVellis (2011). Four of the five scales were placed within this category of the scales being respectable. In Table 2, the descriptive and correlations between the variables used in the model are reported. Fornell and Larker (1981) suggest that discriminant validity can be obtained by comparing the shared variance (squared correlation) between each pair of the construct against the AVEs for the two constructs. It is observed within the constructs, the shared variance observed is lower than the average of their AVEs indicating discriminant validity with the exception of one variable COIV – COVA whereby there

Table 1. CFA items, factor loadings, AVE, alpha and CR.

	Factor loadings	AVE	Alpha	CR
Community Involvement (COIV)		0.57	0.8	0.80
I participate in cultural exchanges between local residents and visitors	0.79			
I cooperate with tourism planning and development initiatives	0.70			
I participate in ecotourism-related activities in communities	0.77			
Nature Relatedness (NRB)		0.57	0.9	0.90
Even in the middle of the city, I notice nature around me	0.62			
I am not separate from nature, but a part of nature	0.71			
I take notice of wildlife wherever I am	0.70			
My connection to nature and the environment is a part of my spirituality	0.83			
I am very aware of environmental issues	0.67			
I feel very connected to all living things and the earth	0.83			
My relationship to nature is an important part of who I am	0.87			
Community Value (COVA)		0.60	0.87	0.86
I believe it is important to take an active role in the civic and sustainability affairs of the community in which I live	0.84			
It is important to me to form close ties with others in my community	0.74			
I believe that it is important to give of one's time to community activities.	0.77			
I feel that it is important to serve as a volunteer in my community	0.75			
Future Intentions (SET)		0.82	0.93	0.93
In future, I'm going to support ecotourism	0.92			
In future, I will try to support ecotourism	0.88			
In future, I will plan to support ecotourism	0.91			
Sustainable Behavior (SUB)		0.52	0.71	0.77
Not going to energy-inefficient destinations	0.75			
Travel shorter distances	0.61			
Travel less due to environmental considerations	0.79			

Table 2. Descriptive statistics and correlations.

	Mean	Std Dev	SUB	SUT	COVA	NSEL	COIV
SUB	2.48	0.72					
SET	3.35	0.97	.467**	1			
COVA	3.22	0.88	.534**	.646**	1		
NRB	2.73	0.68	.515**	.551**	.621**	1	
COIV	2.28	0.73	.601**	.606**	.856**	.439**	1

**Correlation is significant at the 0.01 level (2-tailed).

was a very minor difference of 0.006. This may be due to the fact there is a fine line between involvement and values, but we preferred to keep it separate as the ultimate goals of each are different.

Following validation of the measurement model as discussed above the structural equation model was estimated to test the hypotheses. The model fit is a good fit ($\chi^2 = .621$ with df = 1), $p > .05$ (.431). Given below are the model estimates and the fit statistics and recommended values in Table 3. Fit statistics are shown in Table 4. All of the eight hypotheses are supported, as given in Table 5.

The model fit statistics are presented in Table 4 (Baumgartner & Homburg, 1996) indicating a good fit to the data.

Table 3. Model estimates.

	Standardised β	t
SUB→COIV	.51***	9.08
SUB→COVA	−.12*	−2.84
NRB→COIV	.18*	3.12
NRB→COVA	.43***	10.61
NRB→SET	.41***	4.96
COIV→COVA	.77**	22.70
COIV→SET	.39***	3.35
COVA→SET	.23*	2.11

*Significant at the .05 level. **Significant at the .01 level. ***Significant at the .000 level.

Discussion

The aim of this quantitative research was to examine work identity attributes that relate to two pro-environmental behaviours towards Community Involvement and Community Value for supporting future ecotourism led activities. All eight hypotheses were supported: Sustainable tourism behaviour directly and positively associates with community involvement ($\beta - 0.51$, $t = 9.08$, $p <$ 0.01); Sustainable tourist behaviour directly and but negatively associates community value ($\beta = 0.{-}10$, $t = -2.84$, $p < 0.05$); Nature related behaviour directly and positively associates community involvement ($\beta = 0.18$, $t = 3.12$, $p < 0.01$). Similarly, Nature Related Behavior directly and positively influences Community Value ($\beta = 0.43$, $t = 10.61$, $p < 0.01$). The influences of community involvement and community value are also positive, for instance, Community Involvement has a direct and positive influence with Community Value ($\beta = 0.77$, $t = 22.70$, $p < 0.01$).

Table 4. Fit statistics.

Fit statistics (recommended value)	
CMIN/df (≤3)	0.62
RMR (≥0.5)	0.005
GFI (≥0.9)	0.99
AGFI (≥0.8)	0.98
CFI (≥0.9)	1.00
RMSEA (≤0.08)	0.00

Table 5. Supported and unsupported hypotheses.

	Standardised β	t
SUB→COIV	.51***	9.08
SUB→COVA	−.12*	−2.84
NRB→COIV	.18*	3.12
NRB→COVA	.43***	10.61
NRB→SET	.41***	4.96
COIV→COVA	.77**	22.70
COIV→SET	.39***	3.35
COVA→SET	.23*	2.11
H1	Sustainable Tourist Behavior have a direct positive influence on Community Involvement ($\beta = 0.51$, $t = 9.08$, $p < 0.01$)	Supported
H2	Sustainable Tourist Behavior have a direct negative influence on Community Value ($\beta = 0.{-}12$, $t = -2.84$, $p < 0.05$)	Supported (negative)
H3	Nature Related Behavior directly and positively influences Community Involvement ($\beta = 0.18$, $t = 3.12$, $p < 0.01$).	Supported
H4	Nature Related Behavior directly and positively influences Community Value ($\beta = 0.43$, $t = 10.61$, $p < 0.01$).	Supported
H5	Nature Related Behavior has a direct positive influence with future support for Ecotourism ($\beta = 0.41$, $t = 4.96$, $p <$ 0.01).	Supported
H6	Community Involvement has a direct and positive influence with Community Value ($\beta = 0.77$, $t = 22.70$, $p < 0.01$).	Supported
H7	Community Involvement has a direct and positive relationship with future support for Ecotourism ($\beta = .39$, $t = 35$, $p < 0.05$).	Supported
H8	Community Value has a significant direct and positive relationship with future support for Ecotourism ($\beta = .23$, $t = 2.11$, $p < 0.05$).	Supported

* Significant at the .05 level. ** Significant at the .01 level. *** Significant at the .000 level.

In terms of future support for Ecotourism, this research shows that Nature Related Behavior has a direct positive influence with future support for Ecotourism ($\beta = 0.41$, $t = 4.96$, $p < 0.01$). Furthermore, Community Involvement has a direct and positive relationship with future support for Ecotourism ($\beta = .39$, $t = 35$, $p < 0.05$). As well as, Community Value has a significant direct and positive relationship with future support for Ecotourism ($\beta = .23$, $t = 2.11$, $p < 0.05$). Several authors have incorporated community organisations as a mediating characteristic due to one's belief of being involved in the 'larger scheme of things' (Hughey, Speer, & Peterson, 1999; Sarason, 1993, p. 188), this is parallel to the findings of this study. Although other researchers have found community involvement to be an independent variable (Lee, 2013), this study finds that community involvement mediates behaviour with the intended action. This research is supportive of other findings that greater community participation is an essential ingredient in tourism planning (Keogh, 1990; Murphy, 1988; Simmons, 1994; Simpson, 2008; Tosun, 2006). Further, the results of the study, in particular, suggests that pro-environmental behaviours remain a significant contributor to community involvement and provides future support for sustainable tourism led activities. This is an important contribution and should be tested in future studies to examine pro-environmental behaviour in community participants with future performance in support of sustainable tourism.

Overall this research has shown support for the model with the exception of community values not found to be significant for the future support of sustainable tourism led activities. Sustainable tourism behaviour positively and insignificantly associates future support for sustainable tourism through COVA, and even though Burroughs and Rindfleisch (2002) research on community values seem to have a reasonable degree of predictive validity, this research did not find community values to be significant for future sustainable behaviours which opens avenues for further investigation. Sustainable tourism behaviour indirectly and positively associates future support for sustainable tourism through COIV was also not supported.

Sustainable Tourism behaviour has a significant negative relationship with community values. The greater the behavioural intensity in sustainable tourism, the lesser is the community values, for example, taking an instant or active role in the civic and sustainability affairs of the community or forming close ties, giving ones time or serving as a volunteer. However, Community Involvement strongly and positively associates with community values. This is an interesting finding.

Support for tourism planning is recognised by many authors (Choi & Murray, 2010; Nunkoo & Ramkissoon, 2011; Oviedo-García, Castellano-Verdugo, & Martín-Ruiz, 2008) and it is clear that Community should form an integral part of sustainable development. Community residents should need to consider in assessing the role of the type of participants and be the voice to make a difference. A more plausible explanation might be that nature relatedness is a key qualification in residents aiming to participate in communities and support future sustainable initiatives, as it also directly and positively influences support for future sustainable initiatives. While Sustainable tourism behaviour did not show any relationship with support for sustainable tourism, it was also found to be significant with community involvement.

As Nisbet and Zelenski (2011) propose, policy makers should inspire contact with nature, rather than encouraging people to behave in environmentally friendly ways through fear, guilt or economic incentives. Yet, the main challenge remains in aligning goals of sustainability with the community and work towards improving sustainable tourism. Community support is an antecedent to political support (Vincent & Thompson, 2002). The findings align themselves with themes of other research whereby several studies have shown the importance of examining perceptions, values and interests from community resident to develop strategies for tourism (Vincent & Thompson, 2002).

Segmentation and cluster analysis

The next section identifies the type of cluster groups in a community that would support ecotourism led activities. Residents' perception of tourism, personal values and related constructs were emphasised more importantly than demographic factors (Williams & Lawson, 2001). However, research has shown that positive intention to act in supporting future tourism development was related to gender, age, education, life cycle, length of residence and tourism business connections (Jackson & Inbakaran, 2006). It was also found that those who are involved in proactive community groups strongly supported future tourism development (Jackson & Inbakaran, 2006). While knowing the target group is highly essential (Blamey & Braithwaite, 1997), the same goes with understanding homogenous segments and the type of respondents that would add value to supporting community groups. There are positive associations on the impact of demographic variables and local political community participation (Nowak, Rickson, Ramsey, & Goudy, 1982; Steinberger, 1984). Similarly, McLeod et al. (1996) found positive associations but with different types of participation and demographic variables such as age, gender, education and income. Thus, we propose a two-step cluster analysis to identify the segments that support ecotourism. Market segmentation is a widely accepted tool and identifies the specific needs of homogeneous consumer groups (Dolničar, 2004).

Two-step cluster analysis uses a log-likelihood ratio measure, it is an exploratory instrument developed to show natural groupings or clusters and is used within diverse disciplines for segmentation of demographics and other variables (Okazaki, 2008). Cluster analysis divides the number of respondents into groups as per a similar criterion (Dolnicar, 2002). This method was feasible for this study as it forms clusters based on continuous and categorical data (Norusis, 2008). Researchers use factor analysis to reproduce items and subsequently use cluster analysis (Tkaczynski, Rundle-Thiele, & Beaumont, 2010; Kibicho, 2008).

The first step groups original cases into pre-clusters and uses the raw data in hierarchical clustering as the second step. In step two, clusters are grouped to produce a range of solutions which is further reduced to the best cluster numbers based on the Schwarz's

Bayesian inference criterion (BIC) (Norusis, 2008). BIC is recommended as it avoids the arbitrariness in customary clustering techniques (Norusis, 2008).

We extracted the support for community variables used in this study which was taken from Lee (2013) and the importance of Community Values taken from Burroughs and Rindfleisch (2002), and used Hierarchical clustering to group clusters. The Hierarchical method is suitable if the sample size is moderate (Hair et al., 2006). The clustering size for this research is suitable as per the following variables selected (Dolnicar & Leisch, 2004; Formann, 1984).

(1) I support the development of community-based ecotourism initiatives;
(2) I believe it is important to take an active role in the civic and sustainability affairs of the community in which I live.
(3) I feel that it is important to serve as a volunteer in my community
(4) It is important to me to form close ties with others in my community
(5) I believe that it is important to give one's time to community activities

A further two-step cluster test was taken to evaluate the residents differing in their beliefs towards the importance of taking an active role in the civic and sustainability affairs of a community with two attitudinal variables (a) their concerns towards sustainability tourism and (b) whether they can make a difference to solving sustainability issues as an individual.

Schwarz's Bayesian Criterion (BIC coefficient) showed four clusters as follows (see Table 6):

- Cluster 1 was the smallest cluster and had around 35 respondents (13.2%), followed by,
- Cluster 4 which had 54 respondents (20.3%).
- Cluster 2 had 88 respondents (33.1%) and
- Cluster 3 had 89 respondents (33.5%). Together cluster 2 and 3 had equally the largest number of respondents.

Segmentation discussion

Cluster 1 is similar to cluster 4 that comprises of the younger age group 25–34 (31%), who are employed part-time (27%) and earn in the range of $25,000–$49,000, with most of them having completed a TAFE/Diploma (39%). This is the active group, who is more than likely to support ecotourism development. They believe it is important to take an active role in the civic and sustainability affairs of the community. They find it important to serve as volunteers, important to give one's time to community activities, form close ties with others in their community, cooperate with tourism planning and development initiatives and also support community-based ecotourism initiatives.

Cluster 2 is also similar to cluster one. Most of this cluster falls in the age range of 35–44 (34%), are full time employed (61%); earn in the range of $50,000–$69,000 and hold a TAFE/Diploma qualification (28%). They hold similar perspectives in terms of support for tourism-related sustainable initiatives to cluster 1 but

Table 6. Cluster groups.

Cluster 1	Cluster 2
Age 25-35 (31%) Part Time Employment (27%) $25,000 $49,000 (56.2%) TAFE/Diploma (39%)	Age 35-44 (34%) Full Time Employment(61%) $50,000 $69,000 (31%) TAFE/Diploma (28%)
Cluster 4	**Cluster 3**
Age 25-34 (28%) Full Time Employment(46%) $50,000 $74,000 (43%) University Degree (34.3%)	Age 65+ (96%) Retired(87%) $25,000 $49,000 (46%) High School (54%)

moderately agreed to cooperate with tourism planning and development initiatives.

Cluster 3 comprises mainly of retirees, around 96% of the sample of which 87% are retired. More than half of them have completed their high school (54%). This group believes that it is important to take an active role in the civic and sustainability affairs of the community, important to serve as a volunteer, give time to community activities and form close ties with others in their community. However, there were some restraints on supporting the development of community-based ecotourism initiatives and cooperating with tourism planning and development.

Cluster 4 is an interesting cluster, while they belonged to a similar age group of cluster 1, i.e. most of the respondents fell within the same age group (25–35, 28%), however, these respondents earned higher than cluster 1 ($50,000–$74,000, 43%) and many had acquired an university degree (34%). Overall, they did not agree on the importance of taking an active role in the civic and sustainability affairs of the community, or serving as a volunteer, nor supporting the development of community-based ecotourism initiatives. They also were not willing to give their time to community activities or forming close ties with others in the community. This is a cluster that community leaders would have to avoid.

The analysis showed a good cluster quality. Three clusters were formed: Cluster 1 represented 48% (128) of the respondents that moderately agreed, Cluster 2–14% (38) poorly agreed and Cluster 3–38% (100), highly agreed. ANOVA was used to test the difference and the groups were found to be significantly different. These findings support the views of other authors who have specifically identified an environmentally conscious community and sustainable economic support (Vincent & Thompson, 2002, p. 159).

In sum, only 14% of the sample poorly agreed in taking an active role in the civic and sustainability affairs of a community which is a positive sign of Community Involvement for sustainability tourism to making a difference in solving sustainability issues. Similarly, those aged within 25–35 and earning a higher income and qualified with a university degree were surprisingly not willing to participate nor advance or support ecotourism development and groups falling within this characteristics should potentially be avoided. They could be holding high pressured jobs and perhaps do not have the time. Nonetheless, all categories should be considered on a case by case basis.

Theoretical and organisational implications

The study advances managerial theory and organisation by exploring the role of pro-environmental residents through the qualitative study and achieves a series of key themes, some of which were examined through the quantitative study. Theoretical linkages between behaviours influence community participation and outcomes for fostering ecotourism within community organizations are found to be significant.

In terms of managerial implications, the qualitative research indicated, while knowledge in other areas of sustainability are high, there is a lack of understanding and knowledge on sustainable related tourism issues in terms of work identity attributes. Strategies that include work identity attributes will enhance community organisations to perform more effectively, either independently or using community capacity building. There is a need to identify strategies for degradation of heritage-listed structures through developing strategic plans and action plans for ecotourism welfare in general.

Based on the findings of this research we suggest adopting more effective planning for community capacity building through using work identity attributes for community organisations. Evaluate options and establish stronger community profiling strategies to engage residents. More importantly, community organisations need to consider community profiling which is an iterative process and should be integrated with community impact assessment for efficiency in management.

Sustainability for ecotourism

The foremost consideration for ecotourism development is sustainability (Vincent & Thompson, 2002). Earlier studies have questioned whether ecotourism is an early stage of tourism development? (Butler, 1980). If ecotourism is to be sustained it must contribute to the facets of sustainable development, it must be economically viable, environmentally appropriate and socio-culturally acceptable (Wall, 1997). It is also clear that for ecotourism operation to succeed there should be low ecological impact; include stakeholders in the overall planning and operations and be economically sustainable as a business entity (Drumm & Moore, 2005; Parker & Khare, 2005).

This research has shown that involving work identities can assist in contributing to capacity building for long-term sustainable development. The research is supported by other similar studies that show antecedents of local residential support for ecotourism can help to promote ecotourism sustainability (Lee, 2013). Through identifying clusters of individuals based on their demographics, behaviours and perceived individual effectiveness for supporting ecotourism, the research demonstrates that individuals with pro-environmental behaviour can act as agents of change for realising ecotourism and sustainable development goals. The research highlights that work identities require specific personal attributes to influence the change process through ecotourism. By examining these attributes on community values and involvement, this study offers insights for institutional change processes that meet the central tenets of sustainability for ecotourism.

Sustainability is a growing problem and this research demonstrates that sustainability for ecotourism can be achieved through appropriate involvement of various stakeholders such as communities, design of appropriate policies and adequate planning of work identities.

Limitations and future research

Some caution should be observed in interpreting these results and generalisation of the findings due to a relatively limited sample size. Self-reported measures are generally subjective, so caution requires to be observed. Although, the study offers validity and echoes support for the independent variables examined, it controlled for other variables that are likely to influence the dependent variable, further research can explore these variables.

Sustainable development in tourism is not a simple process. It was beyond the capacity of this research to look at the adaptive capacity which would be useful for future research. The adaptive capacity for individual or group's ability to change for the betterment of the environment could be another means of supporting ecotourism sustainability through community organisations. Other factors that involve attitudes (Lai & Nepal, 2006; Lepp, 2008); perceived benefits and costs affect support for tourism development (Gursoy & Kendall, 2006; Gursoy & Rutherford, 2004; Nicholas et al., 2009; Nunkoo & Ramkissoon, 2011), these should be examined from an ecotourism perspective. Given the similarity of Australia to other Western countries, we believe it is appropriate to extrapolate the findings of this study to other Western countries, as has been done by other researchers using single-country studies (e.g. Komp, van Kersbergen & van Tilburg, 2013). These findings will open further avenues for research. This research is also relevant and offers insightful findings for the overarching UN community involvement and sustainable development goals.

Conclusions

This research sheds light on community organisations and how they can shape the landscape for ecotourism sustainability. This study contributes to conceptual, theoretical and empirical understanding of community capacity building within the context of ecotourism. First, a framework on managing community systems are proposed and empirically tested, which includes theoretical perspectives of work identity attributes like pro-environmental behaviours and building community capacity. Second, this research identifies clusters of individuals based on their demographics, behaviours and perceived individual effectiveness for supporting sustainability for ecotourism. Individuals with pro-environmental behaviour can act as agents of change for realising ecotourism and sustainable development goals. Lastly, it highlights that work identities require specific personal attributes to influence the change process. By examining these attributes on community values and involvement, this study offers insights in the institutional change process that meet the central tenets of sustainability.

Disclosure statement

No potential conflict of interest was reported by the authors.

References

Allen, J. B., & Ferrand, J. L. (1999). Environmental locus of control, sympathy, and pro environmental behavior: A test of Geller's actively caring hypothesis. *Environment and Behavior*, *31*, 338–353.

Alonso, J. (1999). *Comportamiento del Consumidor*. Madrid: Spain ESIC Editorial.

Baker, T. L. (1994). *Doing social research* (2nd ed.). New York, NY: McGraw-Hill.

Bamberger, I. (1983). Value systems, strategies and the performance of small and medium-sized firms. *European Small Business Journal, 1*(4), 25–39.

Baumgartner, H., & Homburg, C. (1996). Applications of structural equation modelling in marketing and consumer research: A review. *International Journal of Research in Marketing, 13*(2), 139–161.

Berlin, S. (1997). *Ways we live: Exploring community*. Gabriola Island, BC: New Society Publishers.

Blamey, R. K. (2001). Principles of ecotourism. In D. Weaver (Ed.), *The encyclopedia of ecotourism*. (pp. 5–22). Wallingford: CABI Publishing.

Blamey, R. K., & Braithwaite, V. A. (1997). A social values segmentation of the potential ecotourism market. *Journal of Sustainable Management, 5*(1), 29–45.

Blank, U. (1989). *The community tourism industry imperative*. New York, NY: Venture Publishing.

Bragg, E. A. (1996). Towards ecological self: Deep ecology meets constructionist self-theory. *Journal of Environmental Psychology, 16*, 93–108.

Brown, A. D. (2018). Identities in organization studies. *Organization Studies, 40*, 1–15.

Budeanu, A. (2007). Sustainable tourist behavior: A discussion of opportunities for change. *International Journal of Consumer Studies, 31*(5), 499–508.

Burroughs, J. E., & Rindfleisch, A. (2002). Materialism and well-being: A conflicting values perspective. *Journal of Consumer Research, 29*(3), 348–370.

Butler, R. W. (1980). The concept of a tourist area cycle of evolution: Implications for management of resources. *Canadian Geographer, 24*, 5–12.

Butterfoss, F. D. (2006). Process evaluation for community participation. *Annual Review Public Health, 27*, 323–340.

Choi, H. C., & Murray, I. (2010). Resident attitudes toward sustainable community tourism. *Journal of Sustainable Tourism, 18*(4), 575–594.

Churchill, G. A. Jr. (1979). A paradigm for developing better measures of marketing constructs. *Journal of Marketing Research, 16*(2), 64–73.

Conn, S. (1998). Living in the earth: Eco-psychology, health and psychotherapy. *Humanistic Psychologist, 26*, 179–198.

Davidson, W. B., & Cotter, P. R. (1997). Psychological sense of community and newspaper readership. *Psychological Reports, 80*(2), 659–665.

De Vaus, D. A. (1993). *Surveys in social research* (3rd ed.). London: UCL Press.

DeVellis, R. F. (2011). *Scale development: Theory and applications* (Vol. 26). Thousand Oaks, CA: Sage.

Dolnicar, S. (2002). A review of data-driven market segmentation in tourism. *Journal of Travel & Tourism Marketing, 12*(1), 1–22.

Dolnicar, S., & Leisch, F. (2004). Segmenting markets by bagged clustering. *Australasian Marketing Journal, 12*(2), 51–65.

Dolničar, S. (2004). Beyond "commonsense segmentation": A systematics of segmentation approaches in tourism. *Journal of Travel Research, 42*(3), 244–250.

Drumm, A., & Moore, A. (2005). *Ecotourism development-a manual for conservation planners and managers. Volume I-an introduction to ecotourism planning*. Arlington, VA: The Nature Conservancy.

Dunlap, R. E., & Van Liere, K. D. (1978). The "new environmental paradigm". *The Journal of Environmental Education, 9*(4), 10–19.

Dunlap, R. E., Van Liere, K. D., Mertig, A. G., & Jones, R. E. (2000). New trends in measuring environmental attitudes: measuring endorsement of the new ecological paradigm: a revised NEP scale. *Journal of Social Issues, 56*(3), 425–442.

Durkheim, E. (1964). *The division of labor in society*. New York, NY: Free Press of Glencoe.

Fallon, L. D., & Kriwoken, L. K. (2003). Community involvement in tourism infrastructure: The case of the Strahan Visitor Centre, Tasmania. *Tourism Management, 24*, 289–308.

Feral, C. H. (1998). The connectedness model and optimal development: Is eco-psychology the answer to emotional well-being? *Humanistic Psychologist, 26*, 243–274.

Fisher, A. (2002). *Radical Eco psychology: Psychology in the service of life*. New York, NY: State University of New York Press.

Fligstein, N., & McAdam, D. (2012). *A theory of fields*. Oxford: Oxford University Press.

Formann, A. K. (1984). *Die latent-class-analyses: Einführung in Theorie und Anwendung*. Weinheim: Beltz.

Fornell, C., & Larker, D. (1981). Structural equation modeling and regression: guidelines for research practice. *Journal of Marketing Research, 18*(1), 39–50.

Fowler, F. J., & Mangione, T. W. (1990). *Standardized survey interviewing*. London: Sage Publications.

Fraj, E., & Martinez, E. (2006). Environmental values and lifestyles as determining factors of ecological consumer behavior: An empirical analysis. *Journal of Consumer Marketing, 23*(3), 133–144.

Galaskiewicz, J. (1997). An urban grants economy revisited: Corporate charitable contributions in the Twin Cities, 1979-81, 1987-89. *Administrative Science Quarterly, 42*(3), 445–471.

Gabriel, Y., Gray, D. E., & Goregaokar, H. (2010). Temporary derailment or the end of the line? Managers coping with unemployment at 50. *Organization Studies, 31*(12), 1687–1712.

Giorgi, S., & Palmisano, S. (2017). Sober intoxication: Institutional contradictions and identity work in the everyday life of four religious communities in Italy. *Organization Studies, 38*, 795–819.

Glynn, T. J. (1986). Neighborhood and sense of community. *Journal of Community Psychology, 14*(4), 341–352.

Gursoy, D., & Kendall, K. W. (2006). Hosting mega events e modeling locals' support. *Annals of Tourism Research, 33*(3), 603–623.

Gursoy, D., & Rutherford, D. G. (2004). Host attitudes toward tourism: An improved structural model. *Annals of Tourism Research, 31*(3), 495–516.

Guthrie, D., & McQuarrie, M. (2005). Privatization and the social contract: Corporate welfare and low-income housing in the United States since 1986. *Research in Political Sociology, 14*, 15–50.

Hair, J. J. F., Black, W. C., Babin, B. J., Anderson, R., & Tatham, R. (2006). *Multivariate Data Analysis* (6th ed.). Englewood Cliffs, NJ: Prentice.

Hallett, T. (2010). The myth incarnate: Recoupling processes, turmoil, and inhabited institutions in an urban elementary school. *American Sociological Review, 75*(1), 52–74.

Hares, A., Dickinson, J., & Wilkes, K. (2010). Climate change and the air travel decisions of UK tourists. *Journal of Transport Geography*, *18*(3), 466–473.

Haywood, K. M. (1988). Responsible and responsive tourism planning in the community. *Tourism Management*, *9*(2), 105–118.

Heinze, K. L., Soderstrom, S., & Heinze, J. E. (2016). Translating institutional change to local communities: The role of linking organizations. *Organization Studies*, *37*(8), 1141–1169.

Homer, P. M., & Kahle, L. R. (1988). A structural equation test of the value-attitude-behavior hierarchy. *Journal of Personality and Social Psychology*, *54*(4), 638–646.

Hughey, J., Speer, P. W., & Peterson, N. A. (1999). Sense of community in community organizations: Structure and evidence of validity. *Journal of Community Psychology*, *27*(1), 97–113.

Jackson, M. S., & Inbakaran, R. J. (2006). Evaluating residents' attitudes and intentions to act towards tourism development in regional Victoria, Australia. *International Journal of Tourism Research*, *8*(5), 355–366.

James, P. (2015). Despite the terrors of typologies: The importance of understanding categories of difference and identity. *Interventions: International Journal of Postcolonial Studies*, *17* (2), 174–195.

Joppe, M. (1996). Sustainable community tourism development revisited. *Tourism Management*, *17*(7), 475–479.

Juvan, E., & Dolnicar, S. (2014). The attitude–behavior gap in eco-tourism. *Annals of Tourism Research*, *48*, 76–95.

Keogh, B. (1990). Public participation in community tourism planning. *Annals of Tourism Research*, *17*(3), 449–465.

Keys, N., Thomsen, D. C., & Smith, T. F. (2016). Adaptive capacity and climate change: The role of community opinion leaders. *Local Environment*, *21*(4), 432–450.

Kibicho, W. (2008). Community-based tourism: A factor-cluster segmentation approach. *Journal of Sustainable Tourism*, *16* (2), 211–231.

King, D. A., & Stewart, W. P. (1996). Eco-tourism and commodification: Protecting people and places. *Biodiversity and Conservation*, *5*, 293–230.

Klöckner, C. A., & Blöbaum, A. (2010). A comprehensive action determination model: Toward a broader understanding of ecological behavior using the example of travel mode choice. *Journal of Environmental Psychology*, *30*(4), 574–586.

Kodeih, F., & Greenwood, R. (2014). Responding to institutional complexity: The role of identity. *Organization Studies*, *35*(1), 7–39.

Komp, K., Van Kersbergen, K., & Van Tilburg, T. (2013). Policies for older volunteers: A study of Germany and Italy, 1990–2008. *Journal of Aging Studies*, *27*(4), 443–455.

Lai, P. H., & Nepal, S. K. (2006). Local perspectives of ecotourism development in Tawushan Nature Reserve, Taiwan. *Tourism Management*, *27*(6), 1117–1129.

Lankford, S. V., & Howard, D. R. (1994). Developing a tourism impact attitude scale. *Annals of Tourism Research*, *21*, 121–139.

Lee, T. H. (2013). Influence analysis of community resident support for eco-tourism development. *Tourism Management*, *34*, 37–46.

Lee, W. H., & Moscardo, G. (2005). Understanding the impact of ecotourism resort experiences on tourists' environmental attitudes and behavioural intentions. *Journal of Sustainable Tourism*, *13*(6), 546–565.

Leopold, A. (1949). *A Sand County Almanac: With essays on conservation from Round River*. New York, NY: Ballantine Books.

Lepp, A. (2008). Tourism and dependency: An analysis of Bigodi village, Uganda. *Tourism Management*, *29*(6), 1206–1214.

Leung, A., Zietsma, C., & Peredo, A. M. (2014). Emergent identity work and institutional change: The 'quiet' revolution of Japanese middle-class housewives. *Organization Studies*, *35*, 423–450.

Lok, J. (2010). Institutional logics as identity projects. *Academy of Management Journal*, *53*(6), 1305–1335.

Marquis, C. (2003). The pressure of the past: Network imprinting in intercorporate communities. *Administrative Science Quarterly*, *48*(4), 655–689.

Marquis, C., Glynn, M. A., & Davis, G. F. (2007). Community isomorphism and corporate social action. *Academy of Management Review*, *32*(3), 925–945.

Martens, S., & Spaargaren, G. (2005). The politics of sustainable consumption: The case of the Netherlands. *Sustainability: Science, Practice and Policy*, *1*, 29–42.

Mayer, F. S., & Frantz, C. M. (2004). The connectedness to nature scale: A measure of individuals' feeling in community with nature. *Journal of Environmental Psychology*, *24*(4), 503–515.

McCool, S. F., & Martin, S. R. (1994). Community attachment and attitudes toward tourism development. *Journal of Travel Research*, *32*(3), 29–34.

McIntyre, G., Hetherington, A., & Inskeep, E. (1993). *Eco-tourism development: Guide for local planners*. Madrid: World Tourism Organization.

McKercher, B., Prideaux, B., Cheung, C., & Law, R. (2010). Achieving voluntary reductions in the carbon footprint of tourism and climate change. *Journal of Sustainable Tourism*, *18*(3), 297–317.

McLeod, J. M., Daily, K., Guo, Z., Eveland Jr, W. P., Bayer, J., Yang, S., & Wang, H. (1996). Community integration, local media use and democratic processes. *Communication Research*, *23*, 179–209.

Miles, M. B., & Huberman, A. M. (1994). *Qualitative data analysis: An expanded sourcebook*. Thousand Oaks, CA: Sage Publications.

Miller, G. A. (2003). Consumerism in sustainable tourism: A survey of UK consumers. *Journal of Sustainable Tourism*, *11* (1), 17–39.

Murphy, P. E. (1981, September). Community attitudes to tourism. *International Journal of Tourism Management*, *2*, 189–195.

Murphy, P. E. (1985). *Tourism a community approach*. New York, NY: Methuen.

Murphy, P. E. (1988). Community driven tourism planning. *Tourism Management*, *9*(2), 96–104.

Nicholas, L., Thapa, B., & Ko, Y. (2009). Residents' perspectives of a world heritage site: The Pitons management area, St. Lucia. *Annals of Tourism Research*, *36*(3), 390–412.

Nisbet, E. K., & Zelenski, J. M. (2011). Underestimating nearby nature: Affective forecasting errors obscure the happy path to sustainability. *Psychological Science*, *22*(9), 1101–1106.

Nisbet, E. K., & Zelenski, J. M. (2013). The NR-6: A new brief measure of nature relatedness. *Frontiers in Psychology*, *4*, 813.

Nisbet, E. K., Zelenski, J. M., & Murphy, S. A. (2009). The nature relatedness scale: Linking individuals' connection with nature to environmental concern and behavior. *Environment and Behavior*, *41*(5), 715–740.

Nordlund, A. M., & Garvill, J. (2002). Value structures behind pro-environmental behavior. *Environment and Behavior*, *34*(6), 740–756.

Norusis, M. (2008). *SPSS 16.0 Advanced Statistical Procedures Companion*. New Jersey: Prentice Hall Press.

Nowak, P. J., Rickson, R. E., Ramsey, C. E., & Goudy, W. J. (1982). Community conflict and models of political participation. *Rural Sociology, 47*, 333–348.

Nunkoo, R., & Ramkissoon, H. (2011). Developing a community support model for tourism. *Annals of Tourism Research, 38*(3), 964–988.

Okazaki, E. (2008). A community-based tourism model: its conception and use. *Journal of Sustainable Tourism, 16*(5), 511–529.

Oviedo-García, M. A., Castellano-Verdugo, M., & Martín-Ruiz, D. (2008). Gaining residents' support for tourism and planning. *International Journal of Tourism Research, 10*(2), 95–109.

Parker, S., & Khare, A. (2005). Understanding success factors for ensuring sustainability in ecotourism development in Southern Africa. *Journal of Ecotourism, 4*(1), 32–46.

Pretty, J. (2002). *Agriculture: Reconnecting people, land, and nature*. London: Earthscan.

Reed, M. (1997). Power relations and community-based tourism planning. *Annals of Tourism Research, 24*(3), 566–591.

Richards, G., & Hall, D. (Eds.). (2003). *Tourism and sustainable community development* (Vol. 7). London: Routledge.

Rokeach, M. (1973). *The nature of human values*. New York, NY: Free Press.

Roszak, T. (2001). *The voice of the earth: An exploration of ecopsychology*. Grand Rapids, MI: Phanes Press.

Sarason, S. B. (1993). American psychology, and the needs for transcendence and community. *Journal of Community Psychology, 21*, 185–202.

Saunders, C. D. (2003). The emerging field of conservation psychology. *Human Ecology Review, 10*, 137–149.

Schultz, P. W. (2000). Empathizing with nature: The effects of perspective taking on concern for environmental issues. *Journal of Social Issues, 56*, 391–406.

Schultz, P. W., & Zelezny, L. C. (1998). Values and pro-environmental behavior: A five-country survey. *Journal of Cross-Cultural Psychology, 29*(4), 540–558.

Schwartz, S. H. (1992). Universals in the content and structure of values: Theoretical advances and empirical tests in 20 countries. *Advances in Experimental Social Psychology, 25*, 1–64.

Simmons, D. G. (1994). Community participation in tourism planning. *Tourism Management, 15*(2), 98–108.

Simpson, K. (2001). Strategic planning and community involvement as contributors to sustainable tourism development. *Current Issues in Tourism, 4*(1), 3–41.

Simpson, M. C. (2008). Community benefit tourism initiatives: A conceptual oxymoron? *Tourism Management, 29*(1), 1–18.

Steinberger, P. J. (1984). Urban politics and communality. *Urban Affairs Quarterly, 20*, 4–21.

Taylor, G. (1995). The community approach: Does it really work? *Tourism Management, 16*(7), 487–489.

Thøgersen, J., Haugaard, P., & Olesen, A. (2010). Consumer responses to ecolabels. *European Journal of Marketing, 44* (11/12), 1787–1810.

Tkaczynski, A., Rundle-Thiele, S., & Beaumont, N. (2010). Destination segmentation: a recommended two-step approach. *Journal of Travel Research, 49*(2), 139–152.

Tosun, C. (2006). Expected nature of community participation in tourism development. *Tourism Management, 27*(3), 493–504.

Vincent, V. C., & Thompson, W. (2002). Assessing community support and sustainability for ecotourism development. *Journal of Travel Research, 41*(2), 153–160.

Wall, G. (1997). Is ecotourism sustainable? *Environmental Management, 21*(4), 483–491.

Walsh, K., & Gordon, J. R. (2008). Creating an individual work identity. *Human Resource Management Review, 18*(1), 46–61.

Ward Thompson, C., Aspinall, P., & Montarzino, A. (2008). The childhood factor: Adult visits to green places and the significance of childhood experience. *Environment and Behavior, 40*, 111–143.

Williams, J., & Lawson, R. (2001). Community issues and resident opinions of tourism. *Annals of Tourism Research, 28*(2), 269–290.

Winson, A. (2006). Ecotourism and sustainability in Cuba: Does socialism make a difference? *Journal of Sustainable Tourism, 14*(1), 6–23.

Wood, M. (2002). *Ecotourism: Principles, practices and policies for sustainability*. Burlington, VT: UNEP and The International Ecotourism Society.

Wright, C., Nyberg, D., & Grant, D. (2012). Hippies on the third floor': Climate change, narrative identity and the micro-politics of corporate environmentalism. *Organization Studies, 33*, 1451–1475.

Wrzesniewski, A., & Dutton, J. E. (2001). Crafting a job: Revisioning employees as active crafters of their work. *Academy of Management Review, 26*(2), 179–201.

Zelenski, J. M., & Nisbet, E. K. (2014). Happiness and feeling connected: The distinct role of nature relatedness. *Environment and Behavior, 46*(1), 3–23.

How eco-service quality affects tourist engagement behaviour in ecotourism: eco-consciousness and eco activity-based learning and S-O-R Theory

Indrani Paul and Gobinda Roy

ABSTRACT

This study aims to examine the growing role of eco-consciousness and eco-activity-based learning in the eco-service quality model and their effect on tourist engagement behaviour in ecotourism destinations. Based on the regression and FsQCA analysis on online travel reviews, the result shows a significant impact of eco-service quality on tourist satisfaction, further leading to their engagement behaviour in revisit, recommendation, and learning engagement. Interestingly, the result reveals that tourists' eco-consciousness significantly stimulates their revisit engagement in ecotourism destinations. Furthermore, eco-activity-based learning also significantly moderates the relationship between the eco-tangible dimension (physical facilities provided in the ecolodges or hotels that are safe and sound for the environment) and tourists' satisfaction. Theoretically, this study adds value to ecotourism literature by extending tourist engagement from psychological (satisfaction) to behavioural engagement (revisit, recommendation, learning engagement). Furthermore, using the S-O-R theory, this study highlights the stimuli of tourist satisfaction and subsequent behavioural responses through tourist engagement behaviour in the ecotourism context.

Introduction

The demand for superior service quality is increasing in the hospitality and tourism industry after the pandemic, as tourists need a safe tourism experience (Günaydin, 2022; Roy & Sharma, 2021). This need has attracted much research attention recently, especially in the ecotourism industry (Irani et al., 2022; Li et al., 2021; Silva et al., 2022). Service quality is crucial for any service industry to satisfy customers' increasing expectations and enrich their service experience (Uzir et al., 2021). Hence, high-quality service in ecotourism destinations is essential for tourist satisfaction and novel experiences (Aseres & Sira, 2020). Such experiences will encourage tourists to adopt sustainable tourism practices and engage them in word-of-mouth communication.

While service quality has been a well-researched topic in the traditional tourism industry (Lai et al., 2018; Park & Jeong, 2019; PJ et al., 2023), the in-depth investigation of the hotel's service quality in ecotourism destinations is not well examined despite many challenges and issues faced by ecotourism service providers (Ban & Ramsaran, 2017). Many ecotourism destinations are located in less developed and remote areas (mainly rural or semi-urban) (Hultman et al., 2015; Wu et al., 2022). Moreover, tourists' expectation of service quality in ecotourism destinations requires unique quality dimensions (such as eco-tangibles, eco-learning, sustainable tourism practices, and eco-activity) to make a memorable tour experience (Ban & Ramsaran, 2017). Therefore, investigating the role of eco-service quality (ECOSERV) (Khan, 2003) in ecotourism destinations is necessary to explore how these extended service quality dimensions can enhance tourist satisfaction, leading to engagement behaviour.

Literature in tourism and hospitality research shows that good service quality improves tourist perception and satisfaction towards the destination (Afthanorhan et al., 2019). This fact influences the destination brand image and loyalty towards the tourist place (Akroush et al., 2016; Dedeoğlu, 2019). Service quality in tourism literature majorly examines tourists' satisfaction, loyalty, and revisit intention as the outcome (response) of service quality attributes (Aksu et al., 2022; Ban et al., 2022). However, tourist engagement is the key driver of tourist satisfaction (Rasoolimanesh et al., 2019). Tourist satisfaction also leads to tourist engagement behaviour in terms of revisit intention, positive reviews, and loyalty.

However, prior literature in ecotourism research investigated the role of service quality on tourist behaviour, which majorly focused on purchase intention for eco-friendly products (often called green purchase), green hotel (eco-label certified) preference to stay, and satisfaction from the ecotourism services (Ahmad & Zhang, 2020; Aksu et al., 2022; Ban et al., 2022). Additionally, adequate service quality stimulates tourist loyalty and engagement through revisit and positive word-of-mouth recommendations (Arici et al., 2023; Dedeoğlu, 2019). As per a service quality study from Parasuraman et al. (1985), service quality should be assessed by comparing tourist's expectations and perceptions. In the ecotourism context, customer perception may differ from the traditional tourism experience in terms of eco-tangible facilities (reuse, recycle, environmentally safe services), ecotourism activities (mountaineering), and eco-conscious behaviour to explore nature which may lead to tourist engagement.

For ecotourism destinations, tourist engagements help visitors learn about the local environment, promote conservation efforts, and improve a sense of community among tourists and residents (Paul & Roy, 2023). Previous literature suggests that a satisfied tourist gets engaged through revisit intention, knowledge assimilation, and recreational activities (such as hiking, bird watching, and local cultural programmes) (Villamediana-Pedrosa et al., 2020). Although service quality is a dominating attribute for any service industry to drive customer satisfaction, however with the growing consciousness over the environment and eco-friendly behaviour, service providers are focusing on reducing carbon footprint and providing environmentally friendly services (Chan, 2021). In this regard, along with the conventional service qualities, how eco service quality (including eco tangible) in the emerging ecotourism destination drives tourist satisfaction and engagement behaviour is yet to be known.

Furthermore, the literature shows that eco-activity, eco-learning, eco-friendly practices, and environmental consciousness of tourists determine tourist perception of destination image and visit intention (Adam et al., 2019). Tourists often get motivated to visit ecotourism destination, where ecolodges organise eco activities (bushwalking, bird watching, mountain biking, climbing trees, cycling) to provide a novel experience (Ban & Ramsaran, 2017). Participating in eco-activities helps tourists in eco-learning through knowledge of local culture, environment, and ecological balance. Therefore, eco-activity-based learning motivates tourists to visit ecotourism destinations (Kim Lian Chan & Baum, 2007). However, research needs to be more active in examining how these activities amplify the effect of ecotourism service qualities, especially eco-tangibles (such as practices to reduce water and energy usage, environmentally safe tourism activity, eco-certificate or eco-labelling of hotels), on tourist satisfaction (Ban & Ramsaran, 2017).

Notably, in ecotourism research, most research is limited to exploring eco-service quality's role on tourist visit/revisit intention, satisfaction, and loyalty behaviour (Arici et al., 2023; Dedeoğlu, 2019). Although, Seyfi et al. (2021) have found that tourist engagement enhances tourist satisfaction and behavioural intentions in terms of revisit intention and word-of-mouth recommendation. However, in ecotourism research, a limited study explores how eco-service quality influences tourist satisfaction and drives tourist engagement behaviours in terms of revisiting intention, recommendation, and learning from the tour experience. Hence, it is essential to know how the emerging service quality dimensions influence tourists' overall engagement behaviour in a remote destination.

Lately, sustainable tourism's growing importance motivates tourists to behave environmentally conscious during their travel experience (Chen & Hung, 2016). Prior literature also reveals that consciousness about the environment and ecology influence individuals' behavioural intentions, such as visit intention (Ahmad et al., 2020). Therefore, this is essential to understand how growing consciousness about the environment further amplifies the effect of satisfaction on tourist engagement behaviour. Hence, this study incorporates the moderating role of eco-consciousness in the conceptual framework of this study. Based on these gaps, this study will investigate the following research questions:

Research Question 1: How does the emerging dimension of eco-service quality influence tourist satisfaction in ecotourism destinations?

Research Question 2: How does tourist satisfaction influence their engagement behaviour in ecotourism destinations?

Research Question 3: Does Eco-activity-based learning moderate the relationship between eco-tangibles and satisfaction in ecotourism destinations?

Research Question 4: Does Eco-consciousness moderate the relationship between satisfaction and engagement behaviours (revisit engagement, recommendation engagement, learning engagement)?

Therefore, this study aims to investigate the role of eco-service quality on tourist engagement behaviour in terms of revisit, recommendation, and learning engagement. The study adopted the S-O-R (Stimuli-Organism-Response) theory to address the above research questions and examine the critical role of service quality in tourist engagement behaviour in the ecotourism context. Eco-service quality dimensions are considered as stimuli that evoke the satisfaction of the tourists (Organism), and subsequently satisfaction leads to tourist engagement behaviours as the final response (outcome). The study has adopted a mixed method approach through regression, and Fuzzy set Qualitative Comparative Analysis (FsQCA).

Theoretically, this study adds value to ecotourism literature by identifying the role of emerging factors in service quality to drive tourist engagement behaviour in the ecotourism context. It shows that eco-consciousness intervenes in tourist satisfaction and revisits engagement in ecotourism destination. Moreover, the result gives insight into the moderation role of eco-activity-based learning in enhancing eco-tangibles' effects on tourists' satisfaction. This study adopts a mixed method approach, and rather than taking scale-based variables, we operationalise the variables based on keyword frequency in tourists' reviews (corpus).

Literature review and hypotheses development

Service quality includes five fundamental dimensions: reliability, tangibles, responsiveness, assurance, and empathy Parasuraman et al. (1985). In the ecotourism industry, extant research shows that along with the basic dimensions of service quality, eco-tangibles, eco-learning, and sustainable practices are also crucial in influencing tourist perception (Aseres & Sira, 2020; Ban & Ramsaran, 2017; Khan, 2003). Since tourists are motivated to visit ecotourism places for a unique experience and environmental benefit (Hultman et al., 2015; Nvight, 1996), therefore we have adopted extended service quality dimensions, including eco tangibility, eco-consciousness, eco-activity-based learning to understand tourist's actual behavioural intentions in the ecotourism places.

Stimulus-organism-response theory

The study's conceptual framework is developed based on the S-O-R (Stimuli-response-organism) theory (Mehrabian & Russell, 1974; Su et al., 2020). According to the S-O-R theory, the stimulus is defined as those factors that affect the individual's internal states and can be conceptualised as an influence that stimulates the individual action (Mehrabian & Russell, 1974). Regarding the tourism industry, service quality attributes accelerate customer behaviour in an affective and cognitive process (Kim et al., 2020; Kim & Moon, 2009). Hence, prior tourism studies used the S-O-R model and found that the service quality attribute of hotels and restaurants stimulates positive or negative emotions among customers and impacts their loyalty towards the hotel through their response behaviour (Jani & Han, 2015).

Based on the literature guidance, the current study adopted the S-O-R theory in the ecotourism context to develop and analyse the relationships among eco-service quality dimensions, satisfaction, and engagement behaviours of tourists. In this study, we conceptualised the dimensions of eco-service quality (such as eco-tangibles, assurance, reliability, responsiveness, empathy, and tangibles) as the stimulus (S) that affect the emotional responses of the tourists. Extant literature also suggested the service quality dimensions as stimuli to impact customers' emotional responses (Famiyeh et al., 2018). Tourist satisfaction is conceptualised as Organism (O) because satisfaction portrays customers feeling and perceptions as an intervening process for the final response (Bagozzi, 1986). Tourists' satisfaction can also lead to engagement behaviour such as word-of-mouth (positive or negative), helping other travellers, and blogging (Van Doorn et al., 2010). Therefore, we considered tourists' engagement behaviour as a response (R) in the form of revisit, recommendation, and learning engagement.

Eco service quality and tourist satisfaction

Eco service quality comprises all the attributes of fundamental dimensions of service quality with a particular focus on eco tangibles (reuse, recycling practice), eco learning (cycling), and eco-activities (guided tour, bird watching). For example, in the case of traditional hospitality services, hotels focus on ensuring a satisfied and enjoyable experience for the guest by providing basic dimensions of service quality (SERVQUAL). In contrast, eco-service quality also focuses on reducing the adverse environmental effect and fostering sustainable practices and activities to promote eco friendly behaviour (Ban & Ramsaran, 2017).

Ecotourism often faces challenges to ensure superior service quality because of its remote location and modern resource constraints (Paul & Roy, 2023). Since service quality is the most dominant attribute for any service industry to retain customers, it is crucial to know whether top-class service assurance in ecotourism destinations remains the priority for tourists. Employees' knowledge and courtesy in a hospitality service provider are a building block for developing their trust, loyalty, and a sense of assurance among the tourists (Roy et al., 2023; Sürücü et al., 2019). Customers often feel assured and confident when staff members demonstrate their learned knowledge and friendly behaviour during service delivery (Balinado et al., 2021). Customers feel satisfied when they get a similar kind of assurance of service quality in a remote location.

The hotel's ability to provide the promised service timely and effectively increased customers' reliability towards the service provider (Meesala & Paul, 2018). Reliability often helps build trust between the customers and the service provider (Tabaeeian et al., 2022). When modern resources are not readily available in ecotourism destinations, a promised and reliable service delivery motivates tourists for such a unique experience.

Willingness to help and offer fast service to tourists builds a responsive image of the service provider (Afthanorhan et al., 2019). While ecolodges show responsiveness in their behaviour through prompt service delivery, it accelerates tourist satisfaction (Ahmad et al., 2020). When the service provider or the staff are responsive, the customer feels more valued, which can lead to higher satisfaction.

Based on the above discussion, we propose the following hypotheses:

H1: Assurance in eco service quality is positively related to Satisfaction

H2: Reliability in eco service quality is positively related to Satisfaction

H3: Responsiveness in eco service quality is positively related to satisfaction

In ecotourism destinations, empathy in service attributes helps tourists get a unique experience in a nature-based place where they feel safer and more valued. Kind and attentive service of the hotels and ecolodges improve tourists' experience (Aseres & Sira, 2020). When a service provider exhibits empathy, clients are more likely to feel heard and cared for (Ban et al., 2022). Thereby, empathy can contribute to the development of a more unique and personalised ecotourism experience.

Tangible cues of hotels, such as surroundings, furnishings, and personnel look of an ecolodge or hotel, are essential to customer satisfaction (Meesala & Paul, 2018). A positive overall impression can be formed from the tangible service cues, and the ecotourism experience can be improved with high-quality tangibles (Ahmad & Zhang, 2020). Nowadays, hotels are giving importance to eco-tangible attributes (physical facilities that are safe for the environment) to adopt sustainable practices (Martínez García de Leaniz et al., 2018). This fact shows hotels eco-tangible priority in service quality attribute. Hence, along with tangible service scape, eco-tangible qualities can enhance the service provider's credibility as a responsible and environmentally conscious ecotourism service provider. This fact can lead to higher satisfaction and trustworthiness towards the destination.

Hence, based on the above premise, we formulate the following hypotheses:

H4: Empathy in eco-service quality is positively related to Satisfaction

H5: Tangible aspect of eco-service quality is positively related to Satisfaction

H6: Eco-tangibles aspect of eco-service quality is positively related to Satisfaction

Satisfaction and tourist engagement behaviour

Tourist satisfaction leads to a memorable tourism experience at the ecotourism site, which motivates them to revisit the place (Tabaeeian et al., 2023). In this regard, the service quality of hotels or ecolodges in an ecotourism destination plays a crucial role in tourist satisfaction or dissatisfaction (Thuy et al., 2020). A satisfied tourist shows their behavioural response by being engaged through their tourism experience. Previous studies suggested that customer (tourist) behavioural responses such as word-of-mouth, providing recommendations, knowledge enhancement, revisiting, referral, and sharing experiences in social media work as a form of engagement (Barari et al., 2021; Pansari & Kumar, 2017; Wei et al., 2013). Hence, a satisfied tourist is likely to engage in revisiting the place.

Therefore, we propose,

H7: Satisfaction in eco-service quality is positively related to revisit engagement.

Furthermore, a satisfied tourist gets involved in recommending the hotel or lodge he/ she has stayed in during their tour (Meng et al., 2008). At the same time, many studies have examined the relationship between satisfaction and positive or negative word-of-mouth (Sharma & Nayak, 2019). A satisfied tourist is more likely to revisit the destination and recommend the same (Hui et al., 2007; Roy et al., 2021). In ecotourism destinations, ecolodges and hotels provide unique experiences compared to traditional tourism destinations. Satisfied tourists share their experiences on social media or review platforms and will likely recommend their fellow travellers. Therefore, we assume,

H8: Satisfaction in eco-service quality is positively related to recommendation engagement.

Ecotourism caters to tourist learning about nature and the host community. One of the primary motivations for ecotourism visits is knowledge enhancement or learning engagement about the local community, culture, and nature (Wen et al., 2019). Although, Tsaur et al. (2010) argued that tourist knowledge and skill improve tourist satisfaction and information search behaviour. In ecotourism destinations, since ecotourists are likely to learn new experiences from the visit, they intend to engage in knowledge enhancement and learn sustainable tourism practices. Therefore, based on the above discussion, we propose the following hypothesis:

H9: Satisfaction in eco-service quality is positively related to learning engagement.

Moderating role of eco-consciousness

Eco-consciousness refers to the state of mind where an individual is aware of their effects on the environment and knows their role as a human on the earth (Martínez García de Leaniz et al., 2018). In this context, eco-consciousness describes the degree to which tourists of ecotourism places are concerned about the environment and ready to take initiatives for environmental conservation by adopting sustainable tourism practices and pro-environmental behaviour (Hameed et al., 2019). Prior studies confirmed the relevant moderating role of eco-consciousness between the brand image and behavioural intentions, such as the intention to visit and recommendation intention (Namkung & Jang, 2013). Peng et al. (2023) suggested that consciousness is a significant moderator between happiness and revisit intention. Hence, we assume that,

H10: Eco-consciousness positively enhances the relationship between satisfaction and revisit engagement

Moreover, eco-consciousness enhances the ecotourism experience. Tourists with higher environmental consciousness show a higher revisit intention to the ecotourism destinations (Huang & Liu, 2017). However, ecotourism places being located in less developed areas, positive word-of-mouth, and recommendations from fellow travellers encourage other eco-tourists to visit the place or the eco-lodges in that place and engage with local activities (Mariani & Borghi, 2022). Therefore, we propose,

H11: Eco-consciousness positively enhances the relationship between satisfaction and recommendation engagement.

Consumers with a higher eco-consciousness tend to formulate positive environmental attitudes (Ahmad et al., 2020). Thereby, tourists' consciousness about the environment fosters their learning engagement from the tour experience (Ballantyne & Packer, 2011). Tourists who are more eco-conscious are motivated to learn more about the destination's history and sustainable practices through the tourist lodge or hotels (Hameed et al., 2022).

Therefore, we propose the following hypothesis:

H12: Eco-consciousness positively enhances the relationship between satisfaction and learning engagement.

Moderating role of eco-activity-based learning

Tourists learning from ecotourism experience is the primary motivating factor found in almost all contexts of ecotourism (Walter, 2013). Ecotourism destinations have the potential to attract tourists through fun learning activities which take care of tourist satisfaction and environmental conservation efforts (Choo & Jamal, 2009; Paul & Roy, 2023). Similarly, eco-travel activities directly impact the environmentally responsible behaviour of the tourists engaged in ecotourism (Chai-Arayalert, 2020; Chiu et al., 2014). Learning activities in ecotourism places or eco-tangible activities such as wildlife watching, interpretive programmes, and natural sites lead to learning outcomes such as knowledge enrichment, conservation effort, environmental awareness, and learning aesthetic appreciation (Walter, 2013). Hence, activities catering to visitors learning during the tour enhance visitors' experience by managing information and adopting environmentally responsible behaviour (Chai-Arayalert, 2020). Furthermore, the eco-tangibles dimension of hotel services promotes environmentally friendly services, intensifying the overall tourism experience in ecotourism places (Ban & Ramsaran, 2017). Therefore, we hypothesise that,

H13: Eco-activity-based learning moderates the relationship between eco-tangibles and satisfaction.

The conceptual model shows the hypothesised relationships (See Figure 1).

Methodology

Study area and data collection

In this study, we collected reviews from an online travel review website of the tourists who have experienced the service in the ecotourism place. For this purpose, we have scrapped the tourist review data from the website of TripAdvisor (http://www. tripadvisor.com) in India during November 2022. TripAdvisor is one of the most extensive review websites worldwide (Xu et al., 2020). Moreover, prior tourism research studies have used TripAdvisor reviews for empirical investigation (Taecharungroj & Mathayomchan, 2019).

We have tourist review data from the hotels, eco-lodges, and homestays in popular ecotourism destinations. The location of the hotels or ecolodges is primarily from the north-eastern area of India, which includes Darjeeling, Gangtok, and their nearby place of ecotourism importance such as Jalpaiguri, Siliguri,

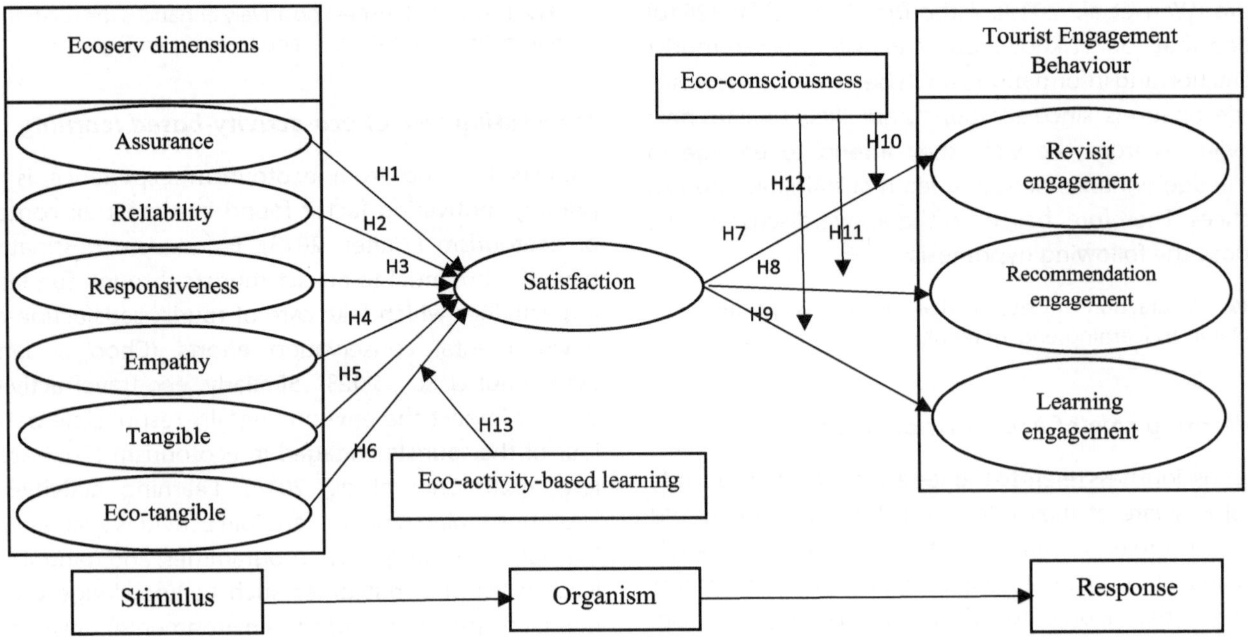

Figure 1. Conceptual model.

Jaldapara, Hasimara, and Dooars. These locations are popular among tourists for providing a nature-based tourism experience (Kumari et al., 2010). Furthermore, ecotourism destination is characterised as a nature-based tourism destination, which promotes sustainability and conservation, and provides educated learning about the environment (Buckley, 1994; Paul & Roy, 2023). Therefore, we have selected the above locations based on the characteristics given in the literature. These locations are nature-based destinations and support conservation.

Initially, we selected 107 hotels from the above locations, as listed on Tripadvisor's website, based on location choice. After careful observation, we have found 81 relevant hotels, ecolodges, and homestays based on the ecotourism location. Finally, we have selected 60 hotels with a minimum of 20 reviews (Maas & Hox, 2005). We have collected review information (review content, hotel star rating) for the first 100–120 reviews based on the recency of the post from each hotel (Roy, 2023). We obtained around 6450 reviews from all 60 hotels (Figure 2).

Data coding and construct operationalisation

We have preprocessed the collected reviews before coding. All the review data have been cleaned and processed into textual data. To clean data before regression analysis, we have removed the stop words, stemming, lemmatising, and converting upper case into lower case using the R Text Mining package (Uysal & Gunal, 2014).

We identified keywords based on the literature review, scale data and conceptual construct, representing key features of SERVQUAL. We mapped the same and related keywords in the corpus for the theoretical specification of the construct (Li et al., 2021). Next, we created a table for the set of keywords for each construct (See Annexure). Following Houston (2004), to measure construct validity, we assessed the ability of the keywords to measure the construct using inter-coder reliability and found kappa score (k) of 0.78. Thus, we established the congruency of keywords between the two coders, representing the key dimensions of eco-service quality (Burla et al., 2008). Finally, we took industry expert opinion to confirm the external validity of the keywords to represent the generalisability (Wolf, 2011).

To operationalise the constructs, first, we have calculated the frequency of each keyword representing a certain construct (for each hotel) in the hypothesised model. For the final statistical counting of each construct, we have taken the cumulative frequency of all the keywords for that construct (Roy & Sharma, 2021). For example, to operationalise the responsiveness, we identified all the keywords with related frequencies and calculated the total count of all those keyword frequencies for count vectorisation (Cui et al., 2022; Roy, 2023) (See Figure 3).

Empirical model specification

To test the hypotheses of the conceptual model, we have segregated the conceptual model into different sub-models to conduct the regression. Model 1 shows the influence of the fundamental service quality dimensions on the satisfaction variable. In contrast, model 2 (including eco-tangibles with the basic model of service quality)

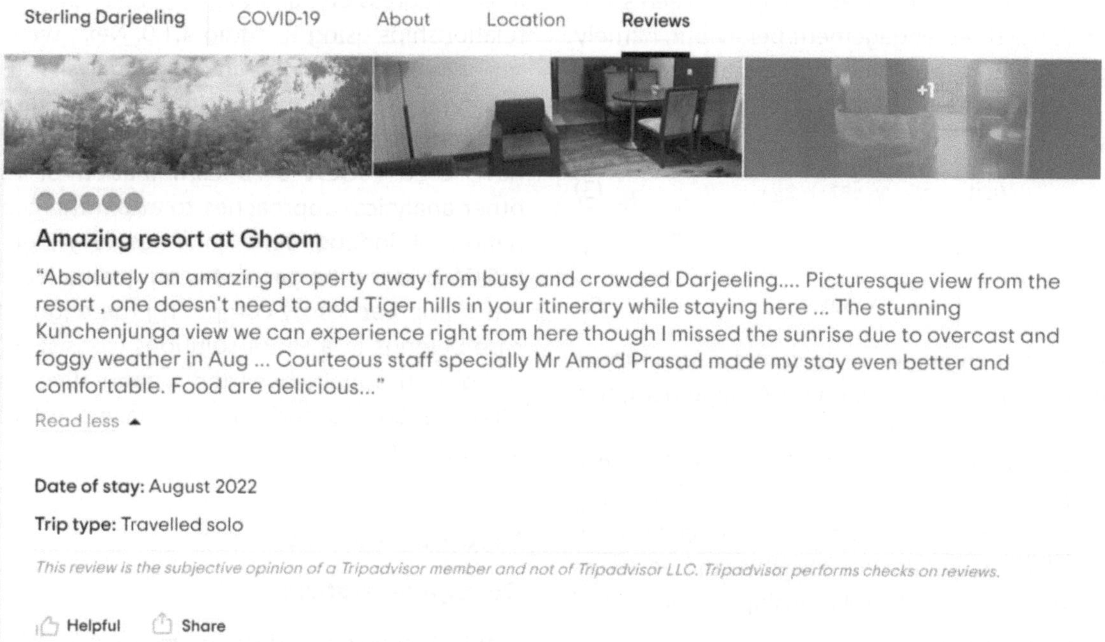

Figure 2. Customer's review on TripAdvisor.

shows the influence of the eco-service quality dimensions on the satisfaction variable. The purpose of this study is to check the incremental role of eco-tangible dimension to explain satisfaction in the overall conceptual model.

In model 1, following equation represents the effect of basic service quality dimensions on satisfaction:

$$Sat = \alpha_0 + \alpha_1 Assu + \alpha_2 Reli + \alpha_3 Resp + \alpha_4 Empa + \alpha_5 Tang + \varepsilon_1 \quad (1)$$

Here, Sat = Satisfaction, Assu = Assurance, Reli = Reliability, Empa = Empathy, Resp = Responsiveness, Tang = Tangible, $\boldsymbol{\varepsilon_1}$ = error term$_1$

In model 2, we incorporate an additional dimension called eco-tangibles. Therefore, in the second model, we examined the impact of eco-service quality on tourist satisfaction. We formulate the relationship by the following equation,

$$Sat = \beta_0 + \beta_1 Assu + \beta_2 Reli + \beta_3 Resp + \beta_4 Empa + \beta_5 Tang + \beta_6 EcoTang + \varepsilon_2 \quad (2)$$

Here, EcoTang = Eco Tangible, $\boldsymbol{\varepsilon_2}$ = error term$_2$

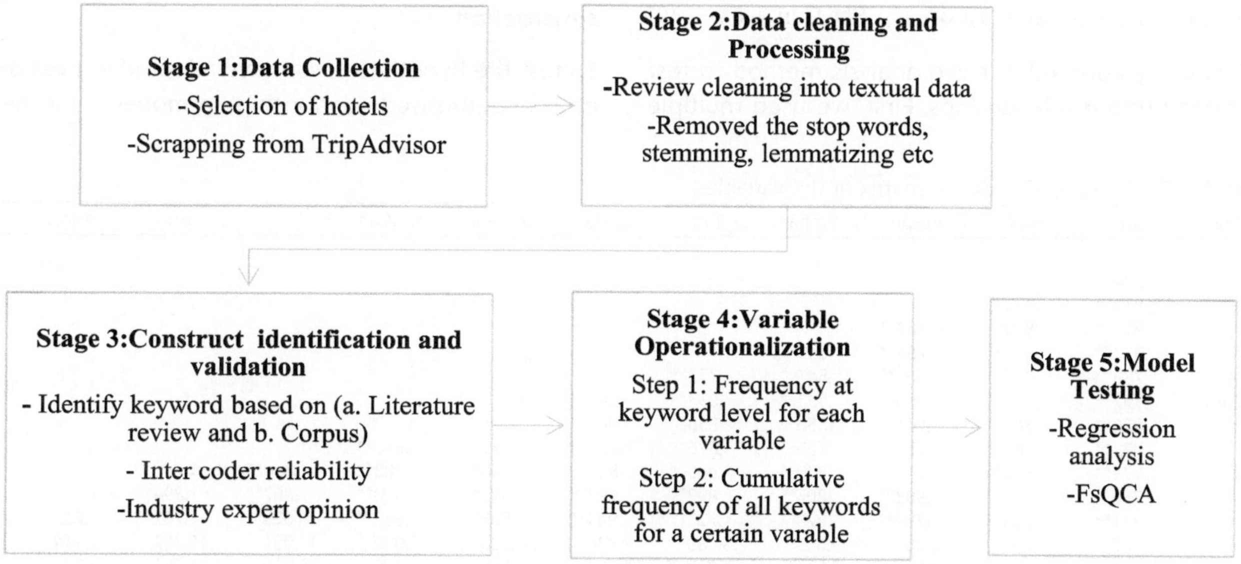

Figure 3. Data processing and analysis flow.

After that, to examine the relationship between satisfaction and all the three-engagement behaviour, namely revisit engagement, recommendation engagement, and learning engagement, we propose the following equations:

$$\text{Visit} = \gamma_{0+}\gamma_1 \text{Sat} + \varepsilon_3 \quad (3)$$

$$\text{Recomm} = \theta_{0+}\theta_1 \text{Sat} + \varepsilon_4 \quad (4)$$

$$\text{Learn} = \lambda_{0+}\lambda_1 \text{Sat} + \varepsilon_5 \quad (5)$$

Here, visit = visit engagement, Recomm = recommendation Engagement, Learn = Learning Engagement, Sat = Satisfaction, $\varepsilon_=$ error term.

In the subsequent analysis, we incorporate the moderating variables. In this, we check the moderating impact of eco-activity-based learning on the relationship between satisfaction and the eco-tangibles dimension of service quality. Therefore, the relationship is represented by the following equation:

$$\text{Sat} = \delta_0 + \delta_1 \text{EcoTan} + \delta_2 \text{EcoLearn} + \delta_3 \text{EcoTan\_EcoLearn} + \varepsilon_6 \quad (6)$$

By incorporating eco-consciousness as a moderating variable, we intend to check the intervening role of eco-consciousness on the relationship between satisfaction and revisit engagement, recommendation engagement, and learning engagement, respectively.

Therefore, we formulate the relationship by the following equations,

$$\text{Visit} = \eta_0 + \eta_1 \text{Sat} + \eta_2 \text{Con} + \eta_3 \text{Sat\_Con} + \varepsilon_7 \quad (7)$$

$$\text{Recomm} = \mu_0 + \mu_1 \text{Sat} + \mu_2 \text{Con} + \mu_3 \text{Sat\_Con} + \varepsilon_8 \quad (8)$$

$$\text{Learn} = \rho_0 + \rho_1 \text{Sat} + \rho_2 \text{Con} + \rho_3 \text{Sat\_Con} + \varepsilon_9 \quad (9)$$

This study adopted a mixed analysis method to test the hypothesised relationships. First, we used multiple linear regression analysis to test the hypothetical relationships using R studio 4.1.0. Next, we used Fuzzy-set Qualitative Comparative Analysis (FsQCA) to achieve more profound insight into the result (Pappas & Woodside, 2021). Several studies on tourism research have used FsQCA or a combination of FsQCA with other analytical approaches to attain the causal combination of independent variables (Seyfi et al., 2021). FsQCA bridges the gap between qualitative and quantitative approaches by computing the degree to which a case belongs to a set of solutions (consists of variables) using both qualitative and quantitative assessments (Ragin, 2000). To perform the analysis, we used FsQCA software 4.0.

Result and discussion

Descriptive statistics

Table 1 shows the descriptive statistics of the sample data, which considered the mean, standard deviation, and significant correlations among variables. Moreover, we checked multicollinearity, normality, and constant variance assumption before conducting regression analysis. We found skewness and kurtosis values range from 0.341 to −0.822 and 0.543 to −0.424, respectively, which are within the recommended range (Walia et al., 2021). To ensure independency, we found all the variance inflation factor values < 4, which confirms no multicollinearity issue among the variables (Hair et al., 2010). We have also found the Breusch–Pagan test score of 7.29, which ensures the homoscedasticity of the residue (Thompson, 2011).

Effect of eco service quality dimensions on satisfaction

To test the hypotheses, we have adopted regression on each hypothesised relationship. In model 1, the results

Table 1. SD, Mean and Correlation matrix of the variables.

Variable	Assu	Reli	Respo	Empa	Tan	EcoTan	EcoLearn	EcoCon	Visit	Recom	Know	Sat
Assu	1											
Reli	.794**	1										
Respo	.812**	.823**	1									
Empa	.811**	.819**	.782**	1								
Tan	.827**	.806**	.781**	.845**	1							
EcoTan	.786**	.892**	.808**	.847**	.772**	1						
EcoLearn	.797**	.851**	.800**	.853**	.833**	.845**	1					
EcoCon	.826**	.867**	.801**	.888**	.856**	.901**	.874**	1				
Visit	.772**	.781**	.757**	.766**	.876**	.722**	.753**	.816**	1			
Recom	.772**	.806**	.741**	.839**	.869**	.833**	.846**	.883**	.880**	1		
Know	.756**	.849**	.803**	.807**	.892**	.792**	.812**	.870**	.887**	.894**	1	
Sat	.771**	.880**	.792**	.828**	.723**	.944**	.825**	.857**	.625**	.742**	.725**	1
Mean	17.507	11.002	12.381	9.603	15.483	7.920	6.725	5.002	13.901	19.702	5.974	4.116
SD	4.491	2.289	2.939	2.264	5.883	1.399	1.568	1.179	8.920	7.051	2.392	0.614

Note: (*) denotes the p value where ***$p < 0.001$; **$p < 0.01$; *$p < 0.05$

Table 2. Model 1 and Model 2.

Model	Relationship among construct	*Hypothesis*	*β*-value	t-value	*P* value	Sig	Decision
Model 1	Assurance→Satisfaction	H1	0.035	0.539	0.591	>0.05	Unsupported
	Reliability→ Satisfaction	H2	0.348	5.240	0.000	<0.001	Supported
	Responsiveness → Satisfaction	H3	0.370	4.200	0.000	<0.001	Supported
	Empathy → Satisfaction	H4	0.218	3.182	0.002	<0.01	Supported
	Tangible → Satisfaction	H5	0.362	4.352	0.000	<0.001	Supported
Model 2	Assurance→Satisfaction	H1	0.030	0.568	0.572	>0.05	Unsupported
	Reliability→ Satisfaction	H2	0.209	3.564	0.000	<0.001	Supported
	Responsiveness → Satisfaction	H3	0.229	3.898	0.000	<0.001	Supported
	Empathy → Satisfaction	H4	0.13	2.273	0.027	<0.05	Supported
	Tangible → Satisfaction	H5	0.293	4.320	0.000	<0.001	Supported
	EcoTangibles → Satisfaction	H6	0.344	5.521	0.000	<0.001	Supported

Note: Model 1: Multiple R-squared: 0.938, Adjusted R-squared: 0.933
Model 2: Multiple R-squared: 0.961, Adjusted R-squared: 0.957
Model 1 represents the dimension of service quality and their impact on satisfaction
Model 2 represents the dimension of eco-service quality on satisfaction, which actually includes the eco tangible dimension with the basic dimensions of service quality.

reveal that reliability, responsibility, empathy, and tangible dimensions of service quality positively affect tourist satisfaction level in hotels and ecolodges of ecotourism destination (H2: β_{reli}→Sat = 0.348, $p < 0.001$; H3: β_{Resp}→Sat = 0.370, $p < .001$; H4: β_{Emp}→Sat = 0.218, $p < .01$; H5: β_{Tang}→Sat = 0.362, $p < .001$). This result is partially supported by prior literature in a different study context (Meesala & Paul, 2018). However, assurance has an insignificant relationship with satisfaction.

In model 2, we added eco-tangibles (H6) and tested the relationship between eco-service quality dimensions and satisfaction. Similar to model 1, in model-2, H1 is not supported. A possible explanation could be the nature of the service destination ecotourism provides. Ecolodges or homestays in ecotourism destinations broadly attract tourists for a novel experience in a natural environment. The ecolodges are usually located in a remote or less developed area (Buckley, 1994). Hence, many ecotourism destinations' quality might need to catch up to assurance in service due to the lack of educated staff, safety in outskirt places, and available modern services. Moreover, tourists may prioritise the serene location, natural beauty of the place, and cleanliness rather than top-notch services from eco-service providers.

However, the result shows a better adjusted R squared value (0.957) while adding an eco-tangibles dimension in model 2. Therefore, a significant positive relationship exists between reliability, responsibility, empathy, tangible and eco-tangibles dimensions on satisfaction (H6: $\beta_{EcoTang}$→Sat = 0.344, $p < 0.001$). Therefore, H2, H3, H4, H5, and H6 are accepted (See Table 2). According to the S-O-R model, in this study, the stimulus, such as reliability, responsiveness, empathy, tangible and eco-tangibles dimensions of service quality, influenced the organism (satisfaction).

Effect of satisfaction on engagement behaviour

We have checked the relationship between satisfaction and engagement behaviour outcomes. The result shows that satisfaction has a positive impact on revisit engagement, recommendation engagement, and learning engagement (H7: $\beta_{Revisit}$→Sat = 0.935, $p < .001$; H8: β_{Recomm}→Sat = 0.936, $p < .001$; H9: β_{Learn}→Sat = 0.953, $p < .001$). Therefore, H7, H8, and H9 are accepted (See Table 3). This result is also in line with the prior literature where Altunel and Erkurt (2015) claimed that satisfaction positively affects recommendation and revisit intention in the context of cultural tourism. According to the S-O-R model, this result explained the various response behaviour in the form of engagement outcomes in ecotourism destinations. While satisfaction represents the organism (emotional process), the result shows the significant impact of satisfaction on the response outcome of revisit engagement, recommendation engagement, and learning engagement from the tour experience.

Result of moderation

The result reveals that eco-consciousness positively moderates the relationship only between satisfaction and revisit engagement (See Figure 4) (H10: $\beta_{Sat \times EcoCon}$ →Revisit = 0.297, $p < .001$). Therefore, H10 is

Table 3. Testing effects of organism (satisfaction) on response (Revisit, Recommendation, Learning).

Relationship among construct	*Hypothesis*	*β*-value	t-value	*P* value	Sig	Decision
Satisfaction→Revisit	H7	0.935	20.02	0.000	<0.001	Supported
Satisfaction→Recommendation	H8	0.936	20.16	0.000	<0.001	Supported
Satisfaction→Learning	H9	0.953	23.9	0.000	<0.001	Supported

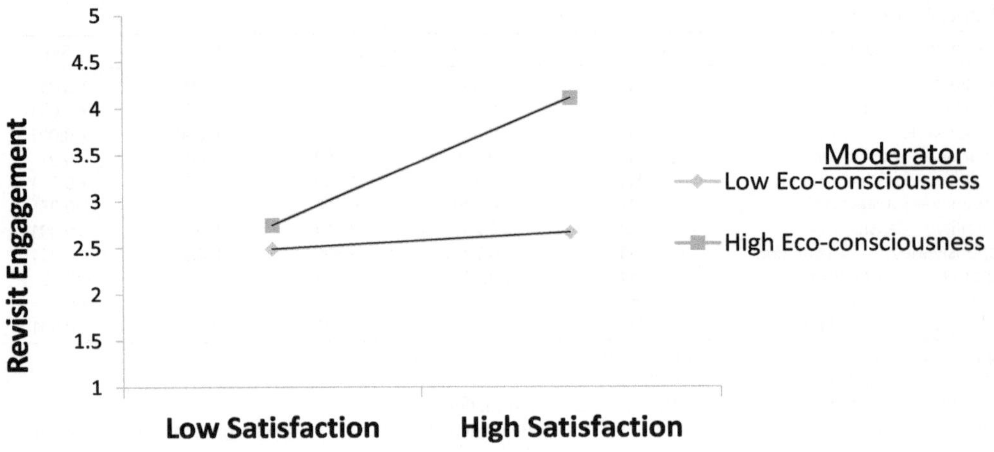

Figure 4. Moderation effect of eco-consciousness.

accepted. Thus, tourist curiosity to explore nature enhances effect of satisfaction on revisit engagement in ecotourism destinations. This result reveals an interesting fact that tourist eco-consciousness behaviour (curiosity, exploration, search) amplifies their satisfaction (Acampora et al., 2022), leading to revisit engagement.

However, the study found an insignificant moderating effect of eco-consciousness between satisfaction and recommendation engagement (H11: $\beta_{\text{Sat x EcoCon}} \rightarrow$Recomm $= 0.001$, $p > .05$) (See Table 4). The probable reason is that ecotourism destinations may provide unique nature-based experiences. Still, tourists have to sacrifice high-end, comfortable tourism experiences (such as luxury hotels and instant access to transportation) in such remote ecotourism places (Liu & Jang, 2009). This fact may result in a lukewarm response in recommending an ecotourism destination to their fellow travellers.

Similarly, the result also shows an insignificant moderating effect of eco-consciousness between satisfaction and learning engagement (H12: $\beta_{\text{Sat x EcoCon}} \rightarrow$Learn $=$ 0.017, $p > .05$). Since travellers often have limited time and resources for knowing the local environment, they may prioritise enjoying the destination rather than devoting all their time to learning-based engagements and activities.

On the other hand, there is a significant moderating effect of eco-activity-based learning between eco-tangibles and satisfaction (H13: $\beta_{\text{EcoTangible x EcoLearn}} \rightarrow$Satisfaction $= 0.331$, $p < .001$) (See Figure 5). In this context, eco-tangible attributes such as recycling, reuse, and environmentally safe events impact tourist satisfaction in ecotourism destinations which is further enhanced by eco-activity-based learning such as guided mountaineering tours (Carvache-Franco et al., 2022).

Result of FsQCA analysis

Data calibration

To transform the raw data into fuzzy data set, FsQCA uses set membership theory (Ragin, 2008). We have used standardised data to perform regression analysis, and the same data set has been transformed into the range of 0–1 (Seyfi et al., 2021). The calibration process considers three specified anchors: a crossover point (0.5), full membership (1), and full non-membership (0) (Ragin, 2008). Next, we found the truth table to determine the configurations or combinations of configurations sufficient to achieve the outcome under consideration (Rubinson, 2019). The following phase involved calculating consistency and coverage for each configuration and identifying enough configurations with coverage larger than 0.2 and consistency greater than 0.8 (Pappas & Woodside, 2021).

Sufficient condition analysis

FsQCA shows a more heterogeneous result as compared to the regression analysis. We found various combinations of sufficient input variables on satisfaction (combination of stimulus) (See Table 5). The result shows three sufficient

Table 4. Moderation testing of eco-consciousness and eco-activity-based learning.

Relationship among construct	*Hypothesis*	β-value	t-value	*P* value	Sig	Decision
Sat x EcoCon→Revisit	H10	0.297	2.422	0.018	<0.05	Supported
Sat x EcoCon→Recomm	H11	0.001	0.042	0.966	>0.05	Unsupported
Sat x EcoCon→Learn	H12	0.017	0.475	0.637	>0.05	Unsupported
EcoTangible **x** EcoLearn→Satisfaction	H13	0.331	4.287	0.000	<0.001	Supported

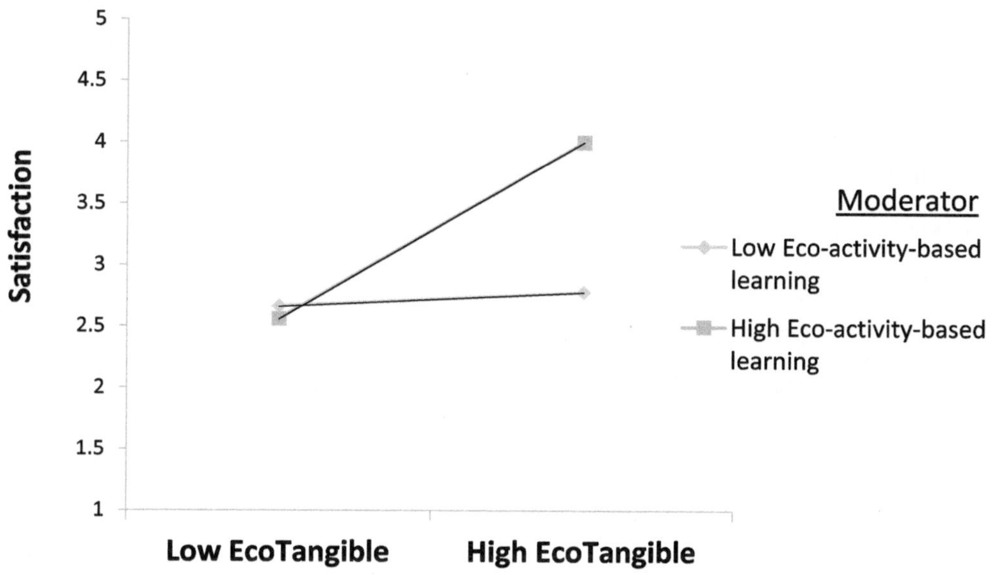

Figure 5. Moderating effect of eco activity-based learning.

configurations (1. ~EcoTang*Tang*Emp*Respons*Reli; 2. ~EcoTang*~Tang*Respons*~Reli; 3. ~EcoTang*Tang*Emp*~Respons*~Reli) to enhance tourist satisfaction. Configuration 1 shows the equal importance of eco tangible, tangible, empathy, responsiveness, and reliability to obtain higher satisfaction. This configuration also supports our regression analysis result (H2, H3, H4, H5, H6).

Configuration 1 shows the highest consistency and unique coverage (0.989 and 0.467, respectively). This result indicated that high eco-tangibles, tangible, empathy, responsiveness, and reliability would bring maximum satisfaction. While configuration 2 explains that high responsiveness and even the lower eco-tangible, tangible, and reliability will also lead to higher satisfaction, the consistency of this proposition is lower (0.76) compared to configuration 1. Finally, configuration 3 shows that high tangible and high empathy will lead to higher satisfaction even though the service might have lower eco-tangible responsiveness and reliability.

Models with consistency over 0.80 are helpful and contribute to theory advancement (Woodside, 2017). Therefore, after obtaining the possible solutions, we ran a test for a specific configuration (configuration 1, which has the highest consistency) and explored how many cases in the sample this configuration (solution) holds true (Figure 6). We have computed the proposition of configuration 1 using fuzzy and function, computed them in a new variable, and renamed the variable as model_1. By selecting the XY Plot option from the FsQCA menu, the new variable (model_1) is plotted against the outcome of interest. It shows that the set of variables in model_1 leads to higher satisfaction.

$$\text{Model\_1} \rightarrow \text{Sat} = f\,(\text{EcoTang*Tang*Emp*Respons*Reli})$$

Necessary condition analysis

In the regression model, we have considered 6 independent variables (dimensions of eco-service quality); therefore, to identify whether any of the conditions are necessary for outcome variable satisfaction, the study analyses whether the condition is always present (or absent) in all cases where the outcome is present (or absent). When the consistency score rises over the 0.9 or 0.8 cut-offs, a condition is considered 'necessary' or 'almost always necessary,' respectively (Pappas & Woodside, 2021). The results of the fsQCA test on the requirement of the criteria concerning satisfaction are shown in

Table 5. Sufficient causal configuration for Satisfaction.

Configurations	Raw Coverage	Unique Coverage	Consistency
Configurations for high satisfaction level			
Sat = ƒ (Assu; Reli; Emp; Respons; Tang; EcoTang)			
EcoTang*Tang*Emp*Respons*Reli	0.772	0.467	0.989
~EcoTang*~Tang*Respons*~Reli	0.4262	0.002	0.762
~EcoTang*Tang*Emp*~Respons*~Reli	0.370	0.001	0.899
Solution Coverage: 0.9566			
Solution Consistency: 0.78719			

Note: Assurance = Assu, Reliability = Reli, Empathy = Emp, Responsiveness = Respons, Tangible = Tang, Eco-tangibles = EcoTang

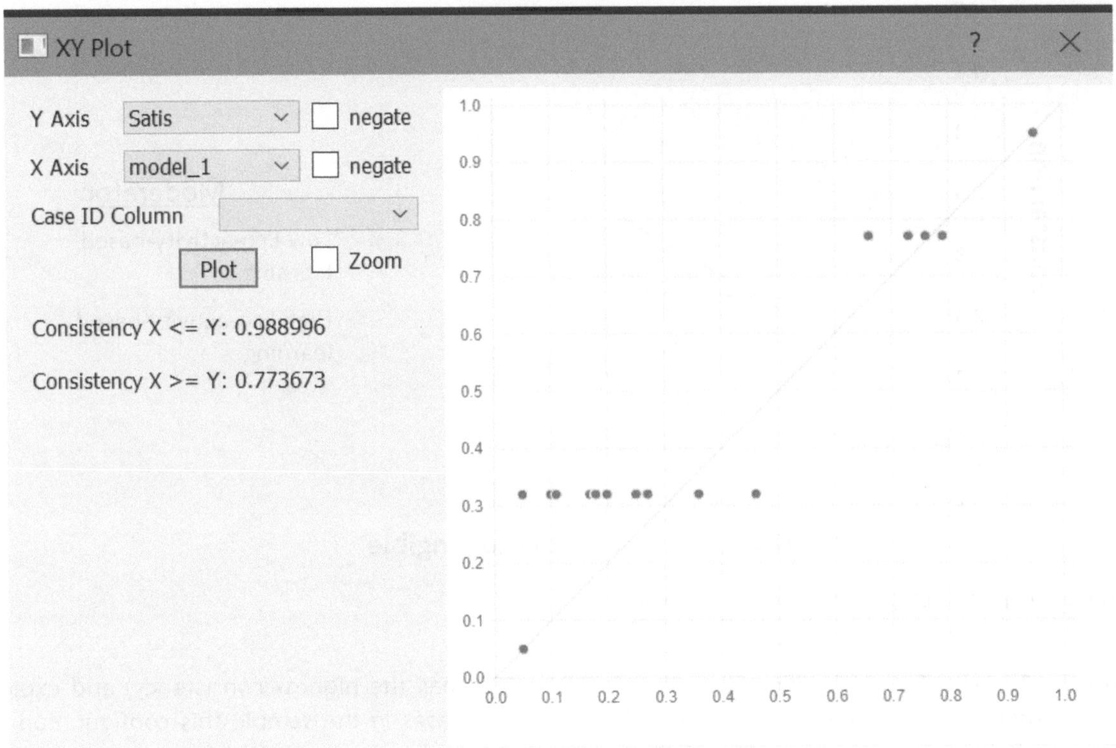

Figure 6. Fuzzy-plot of model_1.

the table (See Table 6). The outcome demonstrates that responsiveness, empathy, assurance, and tangible and eco-tangibles are necessary for satisfaction, while reliability is almost always necessary for satisfaction.

Implication

Theoretical implication

The study has several theoretical implications. Using Stimulus-Organism-Response (S-O-R) framework, the study contributes to unique organism factors such as eco-activity-based learning and eco-consciousness in the eco-service quality model. This insight helps to understand the stimulus of tourist engagement behaviour. The overall framework of S-O-R is explored in the context of eco-service quality to show how each dimension of eco-service quality (S) works as stimuli to boost tourists' engagement (R) in ecotourism as their response behaviour. The study adds value to ecotourism literature by examining tourist engagement behaviour's stimulus, organism, and response in ecolodges or homestays.

Table 6. Analysis of Necessary condition.

Condition	Consistency	Coverage
Assurance	0.914993	0.820785
Reliability	0.881994	0.881362
Responsiveness	0.976327	0.819880
Empathy	0.925036	0.838699
Tangible	0.937590	0.884303
Eco-tangibles	0.929699	0.870091

Methodologically, this study adopts a mixed-method approach. The study used regression analysis as a symmetric analysis approach and FsQCA as an asymmetric analysis approach. Moreover, this contributes to operationalising the new and established constructs by mapping the review of tourists in the online review platform. Instead of taking scale data, we operationalised the constructs based on literature and corpus data and validated the construct reliability by expert opinion.

The study also adds value to ecotourism literature by finding the unique outcome of engagement behaviour, which is influenced by eco-service quality. In the previous literature, service quality in traditional tourism has majorly focused on green buying behaviour, visit intention, loyalty, satisfaction, preference to stay in a homestay, and intention to pay a premium (Hameed et al., 2022; Hultman et al., 2015). Apart from tourist psychological engagement (satisfaction and loyalty), this study highlighted the tourist behavioural response to engagement outcome. Moreover, this study has used a comprehensive service quality dimension using eco-service quality in ecolodges and extended the psychological engagement (satisfaction) and behavioural engagement (revisit, recommendation, Learning).

Since eco-service quality encompasses various dimensions of service quality (e.g. Eco-tangibles, responsiveness, empathy), this uniquely contributes to identifying the tourist's engagement behaviour in ecotourism destinations. Despite the insignificant effect of assurance, our finding showed that eco-service quality influences tourist satisfaction and subsequent engagement behaviour. Eco-tangible factors such as environmentally safe tourism activities, guided tours in protected areas, and bird watching increase tourist satisfaction with the tour experience (Jarvis et al., 2016), enhancing engagement. The study shows that tourists feel satisfied in ecotourism destinations by focusing mainly on eco-activity-based learning and exploring the natural beauty in those places. However, the assurance of getting modern services like traditional tourism destinations is outside their focus to increase satisfaction. Eco-consciousness attributes of tourists instigate tourist curiosity in the ecotourism destination to explore nature and learn about the environment, which gives them satisfaction in the tour experience in ecotourism destination (Mandić & McCool, 2023).

This study uniquely shows the role of eco-activity-based learning in moderating the relationship between eco-tangibles and tourist satisfaction. When tourists learn from the eco-activity in their tour, their satisfaction gets higher from the eco-tangibles' aspect of service quality of the ecolodges. Including eco-activities in ecotourism destinations increase tourist satisfaction and the likelihood of a repeat visit. Tourists learn about the local environment and the efforts to conserve it through eco-activities, including led nature walks, bird viewing, and conservation tours. Also, participating in such activities gives visitors a more immersive experience with the surroundings. Tourists may be improving the environment by participating in eco-activities.

Furthermore, the study identified that eco-consciousness, or an individual's environmental concern, moderates the relationship between tourist satisfaction and revisit intention in ecotourism destinations. Tourists who have high levels of eco-consciousness are likely to place a greater emphasis on the environmental aspects of their travel experiences. They become more satisfied with ecotourism destinations when they are aware of environmental protection and conservation efforts and, therefore, more likely to intend to revisit. Therefore, the eco-consciousness value of a satisfied tourist escalates the intention to revisit the ecotourism destination.

Practical implication

The far-fetching effect of service quality on engagement has several practical implications for ecotourism service providers. Better service quality increases tourist satisfaction and leads to engagement behaviour. Understanding the stimuli of service quality and its effect on satisfaction and engagement will push customer lifetime value (CLV) for the service provider (Kaul, 2017). Therefore, tourists with a high CLV tend to be more loyal and intend to revisit and refer others to the destination.

Second, tourism brands must focus on strategy building around the eco-service quality dimension, especially in ecotourism destinations. Ecolodges might emphasise the eco-tangibles and reliability dimensions among all other service quality attributes. Moreover, ecotourism service providers can position their service by enhancing reliable services with more focus on providing eco-tangibles amenities in remote locations. Such activities may include obtaining eco-labels or eco-certification for hotels, arranging informative guided tours catering to environmental learning, guided mountaineering, and bird watching (Ban & Ramsaran, 2017).

Third, eco-activities offer a unique experience different from traditional tourism activities. This insight can increase the appeal of the destination and make it stand out in the minds of tourists, increasing the likelihood of a return visit. Hence, including eco-activities in ecotourism destinations can lead to a more meaningful and enjoyable tourist experience, increasing satisfaction and the likelihood of repeat visits.

Limitation and future research direction

The study used tourist's online reviews (TripAdvisor) to produce the result. However, future studies could explore with survey data to explore whether the service quality still leads to the same engagement behaviour or not. On the other hand, in this study, the assurance attribute of eco-service quality is insignificant in tourist engagement. Therefore, future research could investigate the role of eco-service quality on engagement behaviour based on large-size survey data. During the data-clearing process, we have not considered the sentiment of reviews based on positive, negative, or neutral sentiment as the primary purpose of data cleaning was to obtain the cumulative frequency of each variable. Therefore, future research can further explore through sentiment analysis how the nature (sentiment) of the review influences the tourist's attitude towards tourist attractions (accommodation, hotel, entertainment).

Conclusion

This study concludes that the eco-service quality of hotels, ecolodges, and homestays in ecotourism destinations significantly impacts tourist satisfaction, leading to various engagement behaviours. The study found the increasing importance of service quality dimensions on tourist satisfaction in ecotourism destinations which emphasises the critical role of eco-tangible, eco-activity-based learning along with the other service quality dimensions. The FsQCA analysis also corroborated the optimal combination of service quality dimensions for tourist satisfaction. We have found the far-reaching effect of tourist satisfaction which influences tourist engagement in terms of revisit engagement, recommendation, and learning about nature.

Moreover, our study finds significant impact of tourist eco-consciousness (explore, discover, curiosity) on tourist engagement behaviour in the form of revisit engagement and this effect does not necessarily lead to recommendation and learning engagement. However, the study reveals that eco-activity-based learning (mountain, riding, bird watching, cycling, guided tour) remarkably enhances the effect of eco-tangibles (recycle, reuse, eco-friendly) on tourist satisfaction in ecotourism destinations.

Disclosure statement

No potential conflict of interest was reported by the author(s).

ORCID

Indrani Paul http://orcid.org/0000-0001-6976-3725
Gobinda Roy http://orcid.org/0000-0001-6660-2846

References

Acampora, A., Lucchetti, M. C., Merli, R., & Ali, F. (2022). The theoretical development and research methodology in green hotels research: A systematic literature review. *Journal of Hospitality and Tourism Management*, *51*, 512–528. https://doi.org/10.1016/j.jhtm.2022.05.007

Adam, I., Adongo, C. A., & Amuquandoh, F. E. (2019). A structural decompositional analysis of eco-visitors' motivations, satisfaction and post-purchase behaviour. *Journal of Ecotourism*, *18*(1), 60–81. https://doi.org/10.1080/14724049.2017.1380657

Afthanorhan, A., Awang, Z., Rashid, N., Foziah, H., & Ghazali, P. (2019). Assessing the effects of service quality on customer satisfaction. *Management Science Letters*, *9*(1), 13–24. https://doi.org/10.5267/j.msl.2018.11.004

Ahmad, W., Kim, W. G., Anwer, Z., & Zhuang, W. (2020). Schwartz personal values, theory of planned behavior and environmental consciousness: How tourists' visiting intentions towards eco-friendly destinations are shaped? *Journal of Business Research*, *110*, 228–236. https://doi.org/10.1016/j.jbusres.2020.01.040

Ahmad, W., & Zhang, Q. (2020). Green purchase intention: Effects of electronic service quality and customer green psychology. *Journal of Cleaner Production*, *267*, 122053. https://doi.org/10.1016/j.jclepro.2020.122053

Akroush, M. N., Jraisat, L. E., Kurdieh, D. J., AL-Faouri, R. N., & Qatu, L. T. (2016). Tourism service quality and destination loyalty–the mediating role of destination image from international tourists' perspectives. *Tourism Review*, *71*(1), 18–44. https://doi.org/10.1108/TR-11-2014-0057

Aksu, A., Albayrak, T., & Caber, M. (2022). Hotel customer segmentation according to eco-service quality perception: The case of Russian tourists. *Journal of Hospitality and Tourism Insights*, *5*(3), 501–514. https://doi.org/10.1108/JHTI-10-2020-0188

Altunel, M. C., & Erkurt, B. (2015). Cultural tourism in Istanbul: The mediation effect of tourist experience and satisfaction on the relationship between involvement and recommendation intention. *Journal of Destination Marketing & Management*, *4*(4), 213–221. https://doi.org/10.1016/j.jdmm.2015.06.003

Arici, H. E., Cakmakoglu Arıcı, N., & Altinay, L. (2023). The use of big data analytics to discover customers' perceptions of and satisfaction with green hotel service quality. *Current Issues in Tourism*, *26*(2), 270–288.

Aseres, S. A., & Sira, R. K. (2020). An exploratory study of ecotourism services quality (ESQ) in Bale Mountains National Park (BMNP), Ethiopia: Using an ECOSERV model. *Annals of Leisure Research*, *23*(3), 386–406. https://doi.org/10.1080/11745398.2019.1642769

Bagozzi, R. P. (1986). *Principles of marketing management*. Science Research Associates.

Balinado, J. R., Prasetyo, Y. T., Young, M. N., Persada, S. F., Miraja, B. A., & Redi, A. A. N. P. (2021). The effect of service quality on customer satisfaction in an automotive after-sales service. *Journal of Open Innovation: Technology, Market, and Complexity*, *7*(2), 116. https://doi.org/10.3390/joitmc7020116

Ballantyne, R., & Packer, J. (2011). Using tourism free-choice learning experiences to promote environmentally sustainable behaviour: The role of post-visit 'action resources'. *Environmental Education Research*, *17*(2), 201–215. https://doi.org/10.1080/13504622.2010.530645

Ban, J., Kim, H. J., Sheehan, B., & Prideaux, B. (2022). How service quality and perceived value affect behavioral intentions of ecolodgeecolodge guests: The moderating effect of prior visit. *Journal of Vacation Marketing*, *28*(2), 244–257. https://doi.org/10.1177/13567667211042641

Ban, J., & Ramsaran, R. R. (2017). An exploratory examination of service quality attributes in the ecotourism industry. *Journal of Travel & Tourism Marketing*, *34*(1), 132–148. https://doi.org/10.1080/10548408.2016.1141155

Barari, M., Ross, M., Thaichon, S., & Surachartkumtonkun, J. (2021). A meta-analysis of customer engagement behaviour. *International Journal of Consumer Studies*, *45*(4), 457–477. https://doi.org/10.1111/ijcs.12609

Buckley, R. (1994). A framework for ecotourism. *Annals of Tourism Research*, *21*(3), 661–665. https://doi.org/10.1016/0160-7383(94)90126-0

Burla, L., Knierim, B., Barth, J., Liewald, K., Duetz, M., & Abel, T. (2008). From text to codings: Intercoder reliability assessment in qualitative content analysis. *Nursing Research*, *57*(2), 113–117. https://doi.org/10.1097/01.NNR.0000313482.33917.7d

Carvache-Franco, M., Pérez-Orozco, A., Carvache-Franco, W., Víquez-Paniagua, A. G., & Carvache-Franco, O. (2022). Motivations and their influence on satisfaction and loyalty in ecotourism: A study of the foreign tourist in Costa Rica. *Anatolia*, *33*(3), 347–361. https://doi.org/10.1080/13032917.2021.1933115

Chai-Arayalert, S. (2020). Smart application of learning ecotourism for young ecotourists. *Cogent Social Sciences*, *6*(1), 1772558. https://doi.org/10.1080/23311886.2020.1772558

Chan, E. S. (2021). Why do hotels find reducing their carbon footprint difficult? *International Journal of Contemporary Hospitality Management*, *33*(5), 1646–1667. https://doi.org/10.1108/IJCHM-10-2020-1151

Chen, S. C., & Hung, C. W. (2016). Elucidating the factors influencing the acceptance of green products: An extension of theory of planned behavior. *Technological Forecasting and Social Change*, *112*, 155–163. https://doi.org/10.1016/j.techfore.2016.08.022

Chiu, Y. T. H., Lee, W. I., & Chen, T. H. (2014). Environmentally responsible behavior in ecotourism: Antecedents and implications. *Tourism Management*, *40*, 321–329. https://doi.org/10.1016/j.tourman.2013.06.013

Choo, H., & Jamal, T. (2009). Tourism on organic farms in South Korea: A new form of ecotourism=. *Journal of Sustainable Tourism*, *17*(4), 431–454. https://doi.org/10.1080/09669580802713440

Cui, Y., Kim, S., & Feng, S. (2022). Exploring success factors of tourism performing arts by analyses of online reviews. *Journal of Hospitality and Tourism Technology*, *14*(1), 37–52. https://doi.org/10.1108/JHTT-05-2021-0140

Dedeoğlu, B. B. (2019). Shaping tourists' destination quality perception and loyalty through destination country image: The importance of involvement and perceived value. *Tourism Management Perspectives*, *29*, 105–117. https://doi.org/10.1016/j.tmp.2018.11.006

Famiyeh, S., Asante-Darko, D., & Kwarteng, A. (2018). Service quality, customer satisfaction, and loyalty in the banking sector: The moderating role of organizational culture. *International Journal of Quality & Reliability Management*, *35*(8), 1546–1567. https://doi.org/10.1108/IJQRM-01-2017-0008

Günaydin, Y. (2022). Service quality in hospitality businesses and its effect on revisit intention during the COVID-19. *Journal of Tourism Theory and Research*, *8*(2), 37–46. https://doi.org/10.24288/jttr.1066010

Hair, J. F., Black, W. C., Babin, B. J., & Anderson, R. E. (2010). *Multivariate data analysis: A global perspective* (7th Edn). Pearson Prentice Hall.

Hameed, I., Hussain, H., & Khan, K. (2022). The role of green practices toward the green word-of-mouth using stimulus-organism-response model. *Journal of Hospitality and Tourism Insights*, *5*(5), 1046–1061. https://doi.org/10.1108/JHTI-04-2021-0096

Hameed, I., Waris, I., & Amin ul Haq, M. (2019). Predicting eco-conscious consumer behavior using theory of planned behavior in Pakistan. *Environmental Science and Pollution Research*, *26*(15), 15535–15547. https://doi.org/10.1007/s11356-019-04967-9

Houston, M. B. (2004). Assessing the validity of secondary data proxies for marketing constructs. *Journal of Business Research*, *57*(2), 154–161. https://doi.org/10.1016/S0148-2963(01)00299-5

Huang, Y. C., & Liu, C. H. S. (2017). Moderating and mediating roles of environmental concern and ecotourism experience for revisit intention. *International Journal of Contemporary Hospitality Management*, *29*(7), 1854–1872. https://doi.org/10.1108/IJCHM-12-2015-0677

Hui, T. K., Wan, D., & Ho, A. (2007). Tourists' satisfaction, recommendation and revisiting Singapore. *Tourism Management*, *28*(4), 965–975. https://doi.org/10.1016/j.tourman.2006.08.008

Hultman, M., Kazeminia, A., & Ghasemi, V. (2015). Intention to visit and willingness to pay premium for ecotourism: The impact of attitude, materialism, and motivation. *Journal of Business Research*, *68*(9), 1854–1861. https://doi.org/10.1016/j.jbusres.2015.01.013

Irani, H. R., Golverdi, M., Rastegar, R., & Rezaei, S. (2022). Rapid ecotourism development and challenges: The case of eco-lodge in Iran. *Tourism Recreation Research*, 1–6. https://doi.org/10.1080/02508281.2022.2086417

Jani, D., & Han, H. (2015). Influence of environmental stimuli on hotel customer emotional loyalty response: Testing the moderating effect of the big five personality factors. *International Journal of Hospitality Management*, *44*, 48–57. https://doi.org/10.1016/j.ijhm.2014.10.006

Jarvis, D., Stoeckl, N., & Liu, H. B. (2016). The impact of economic, social and environmental factors on trip satisfaction and the likelihood of visitors returning. *Tourism Management*, *52*, 1–18. https://doi.org/10.1016/j.tourman.2015.06.003

Kaul, D. (2017). Customer relationship management (CRM), customer satisfaction and customer lifetime value (CLV) in retail. *Review of Professional Management*, *15*(2).

Khan, M. (2003). ECOSERV: Ecotourists' quality expectations. *Annals of Tourism Research*, *30*(1), 109–124. https://doi.org/10.1016/S0160-7383(02)00032-4

Kim Lian Chan, J., & Baum, T. (2007). Motivation factors of eco-tourists in ecolodge accommodation: The push and pull factors. *Asia Pacific Journal of Tourism Research*, *12*(4), 349–364. https://doi.org/10.1080/10941660701761027

Kim, M. J., Lee, C. K., & Jung, T. (2020). Exploring consumer behavior in virtual reality tourism using an extended stimulus-organism-response model. *Journal of Travel Research*, *59*(1), 69–89. https://doi.org/10.1177/0047287518818915

Kim, W. G., & Moon, Y. J. (2009). Customers' cognitive, emotional, and actionable response to the servicescape: A test of the moderating effect of the restaurant type. *International Journal of Hospitality Management*, *28*(1), 144–156. https://doi.org/10.1016/j.ijhm.2008.06.010

Kumari, S., Behera, M. D., & Tewari, H. R. (2010). Identification of potential ecotourism sites in West District, Sikkim using geospatial tools. *Tropical Ecology*, *51*(1), 75–85.

Lai, I. K., Hitchcock, M., Yang, T., & Lu, T. W. (2018). Literature review on service quality in hospitality and tourism (1984-2014): Future directions and trends. *International Journal of Contemporary Hospitality Management*, *30*(1), 114–159. https://doi.org/10.1108/IJCHM-08-2016-0408

Li, T., Liu, F., & Soutar, G. N. (2021). Experiences and value perceptions of an ecotourism trip–an empirical study of outbound Chinese tourists. *Tourism Recreation Research*, *46*(3), 333–344. https://doi.org/10.1080/02508281.2020.1804736

Liu, Y., & Jang, S. S. (2009). Perceptions of Chinese restaurants in the US: What affects customer satisfaction and behavioral intentions? *International Journal of Hospitality Management*, *28*(3), 338–348. https://doi.org/10.1016/j.ijhm.2008.10.008

Maas, C. J., & Hox, J. J. (2005). Sufficient sample sizes for multilevel modeling. *Methodology*, *1*(3), 86–92. https://doi.org/10.1027/1614-2241.1.3.86

Mandić, A., & McCool, S. F. (2023). A critical review and assessment of the last 15 years of experience design research in a nature-based tourism context. *Journal of Ecotourism*, *22*(1), 208–235. https://doi.org/10.1080/14724049.2022.2099877

Mariani, M., & Borghi, M. (2022). Exploring environmental concerns on digital platforms through big data: The effect of online consumers' environmental discourse on online review ratings. *Journal of Sustainable Tourism*, 1–20. https://doi.org/10.1080/09669582.2022.2033982

Martínez García de Leaniz, P., Herrero Crespo, Á, & Gómez López, R. (2018). Customer responses to environmentally certified hotels: The moderating effect of environmental consciousness on the formation of behavioral intentions. *Journal of Sustainable Tourism*, *26*(7), 1160–1177. https://doi.org/10.1080/09669582.2017.1349775

Meesala, A., & Paul, J. (2018). Service quality, consumer satisfaction and loyalty in hospitals: Thinking for the future. *Journal of Retailing and Consumer Services*, *40*, 261–269. https://doi.org/10.1016/j.jretconser.2016.10.011

Mehrabian, A., & Russell, J. A. (1974). *An approach to environmental psychology*. MIT Press.

Meng, F., Tepanon, Y., & Uysal, M. (2008). Measuring tourist satisfaction by attribute and motivation: The case of a nature-based resort. *Journal of Vacation Marketing*, *14*(1), 41–56. https://doi.org/10.1177/1356766707084218

Namkung, Y., & Jang, S. S. (2013). Effects of restaurant green practices on brand equity formation: Do green practices really matter? *International Journal of Hospitality Management*, *33*, 85–95. https://doi.org/10.1016/j.ijhm.2012.06.006

Nvight, P. A. (1996). North American ecotourism markets: Motivations, preferences, and destinations. *Journal of Travel Research*, *35*(1), 3–10. https://doi.org/10.1177/004728759603500102

Pansari, A., & Kumar, V. (2017). Customer engagement: The construct, antecedents, and consequences. *Journal of the Academy of Marketing Science*, *45*(3), 294–311. https://doi.org/10.1007/s11747-016-0485-6

Pappas, I. O., & Woodside, A. G. (2021). Fuzzy-set qualitative comparative analysis (fsQCA): guidelines for research practice in information systems and marketing. *International Journal of Information Management*, *58*, 102310. https://doi.org/10.1016/j.ijinfomgt.2021.102310

Parasuraman, A., Zeithaml, V. A., & Berry, L. L. (1985). A conceptual model of service quality and its implications for future research. *Journal of Marketing*, *49*(4), 41–50. https://doi.org/10.1177/002224298504900403

Park, J., & Jeong, E. (2019). Service quality in tourism: A systematic literature review and keyword network analysis. *Sustainability*, *11*(13), 3665. https://doi.org/10.3390/su11133665

Paul, I., & Roy, G. (2023). Tourist's engagement in ecotourism: A review and research agenda. *Journal of Hospitality and Tourism Management*, *54*, 316–328. https://doi.org/10.1016/j.jhtm.2023.01.002

Peng, J., Yang, X., Fu, S., & Huan, T. C. T. (2023). Exploring the influence of tourists' happiness on revisit intention in the context of Traditional Chinese Medicine cultural tourism. *Tourism Management*, *94*, 104647. https://doi.org/10.1016/j.tourman.2022.104647

PJ, S., Singh, K., Kokkranikal, J., Bharadwaj, R., Rai, S., & Antony, J. (2023). Service quality and customer satisfaction in hospitality, leisure, sport and tourism: An assessment of research in Web of Science. *Journal of Quality Assurance in Hospitality & Tourism*, *24*(1), 24–50. https://doi.org/10.1080/1528008X.2021.2012735

Ragin, C. C. (2000). *Fuzzy-set social science*. University of Chicago Press.

Ragin, C. C. (2008). Measurement Versus Calibration: A Set-Theoretic Approach. In Janet M. Box-Steffensmeier, Henry E. Brady, and David Collier (eds), The Oxford Handbook of Political Methodology (Pp.174–198). Oxford Academic. https://doi.org/10.1093/oxfordhb/9780199286546.003.0008

Rasoolimanesh, S. M., Md Noor, S., Schuberth, F., & Jaafar, M. (2019). Investigating the effects of tourist engagement on satisfaction and loyalty. *The Service Industries Journal*, *39*(7-8), 559–574. https://doi.org/10.1080/02642069.2019.1570152

Roy, G. (2023). Travelers' online review on hotel performance–Analyzing facts with the Theory of Lodging and sentiment analysis. *International Journal of Hospitality Management*, *111*, 103459. https://doi.org/10.1016/j.ijhm.2023.103459

Roy, G., Datta, B., Mukherjee, S., & Basu, R. (2021). Effect of eWOM stimuli and eWOM response on perceived service quality and online recommendation. *Tourism Recreation Research*, *46*(4), 457–472. https://doi.org/10.1080/02508281.2020.1809822

Roy, G., Datta, B., Mukherjee, S., Eckert, A., & Dixit, S. K. (2023). How online travel reviews sources affect travelers' behavioral intentions? Analysis with source credibility theory. *Tourism Planning & Development*, 1–31. https://doi.org/10.1080/21568316.2023.2229296

Roy, G., & Sharma, S. (2021). Analyzing one-day tour trends during COVID-19 disruption–applying push and pull theory and text mining approach. *Tourism Recreation Research*, *46*(2), 288–303. https://doi.org/10.1080/02508281.2020.1858614

Rubinson, C. (2019). Presenting qualitative comparative analysis: Notation, tabular layout, and visualization. *Methodological Innovations*, *12*(2). https://doi.org/10.1177/2059799119862110

Seyfi, S., Rasoolimanesh, S. M., Vafaei-Zadeh, A., & Esfandiar, K. (2021). Can tourist engagement enhance tourist behavioural intentions? A combination of PLS-SEM and fsQCA approaches. *Tourism Recreation Research*, 1–12. https://doi.org/10.1080/02508281.2021.1981092

Sharma, P., & Nayak, J. K. (2019). Understanding memorable tourism experiences as the determinants of tourists' behaviour. *International Journal of Tourism Research*, *21*(4), 504–518. https://doi.org/10.1002/jtr.2278

Silva, A. S., Fialho, R. G. G., da Costa, M. F., & de Oliveira Campos, P. (2022). Antecedents of the intention to visit ecotourism destinations that suffered environmental disasters. *Tourism Management Perspectives*, *41*, 100942. https://doi.org/10.1016/j.tmp.2022.100942

Su, L., Hsu, M. K., & Boostrom Jr, R. E. (2020). From recreation to responsibility: Increasing environmentally responsible behavior in tourism. *Journal of Business Research*, *109*, 557–573. https://doi.org/10.1016/j.jbusres.2018.12.055

Sürücü, Ö, Öztürk, Y., Okumus, F., & Bilgihan, A. (2019). Brand awareness, image, physical quality and employee behavior as building blocks of customer-based brand equity: Consequences in the hotel context. *Journal of Hospitality and Tourism Management*, *40*, 114–124. https://doi.org/10.1016/j.jhtm.2019.07.002

Tabaeeian, R. A., Yazdi, A., Mokhtari, N., & Khoshfetrat, A. (2023). Host-tourist interaction, revisit intention and memorable tourism experience through relationship quality and perceived service quality in ecotourism. *Journal of Ecotourism*, *22*(3), 406–429.

Taecharungroj, V., & Mathayomchan, B. (2019). Analysing TripAdvisor reviews of tourist attractions in Phuket, Thailand. *Tourism Management*, *75*, 550–568. https://doi.org/10.1016/j.tourman.2019.06.020

Thompson, A. (2011). Terrorism and tourism in developed versus developing countries. *Tourism Economics*, *17*(3), 693–700. https://doi.org/10.5367/te.2011.0064

Thuy, V. T. N., Tuu, H. H., Llosa, S., & Thao, H. D. P. (2020). An application of Tetraclass model for evaluating ecotourism service quality in Vietnam. *Tourism and Hospitality Research*, *20*(3), 300–316. https://doi.org/10.1177/1467358419867741

Tsaur, S. H., Yen, C. H., & Chen, C. L. (2010). Independent tourist knowledge and skills. *Annals of Tourism Research*, *37*(4), 1035–1054. https://doi.org/10.1016/j.annals.2010.04.001

Uysal, A. K., & Gunal, S. (2014). The impact of preprocessing on text classification. *Information Processing & Management*, *50*(1), 104–112. https://doi.org/10.1016/j.ipm.2013.08.006

Uzir, M. U. H., Al Halbusi, H., Thurasamy, R., Hock, R. L. T., Aljaberi, M. A., Hasan, N., & Hamid, M. (2021). The effects of service quality, perceived value and trust in home delivery service personnel on customer satisfaction: Evidence from a developing country. *Journal of Retailing and Consumer Services*, *63*, 102721. https://doi.org/10.1016/j.jretconser.2021.102721

Van Doorn, J., Lemon, K. N., Mittal, V., Nass, S., Pick, D., Pirner, P., & Verhoef, P. C. (2010). Customer engagement behavior: Theoretical foundations and research directions. *Journal of Service Research*, *13*(3), 253–266. https://doi.org/10.1177/1094670510375599

Villamediana-Pedrosa, J. D., Vila-López, N., & Küster-Boluda, I. (2020). Predictors of tourist engagement: Travel motives and tourism destination profiles. *Journal of Destination Marketing & Management*, *16*, 100412. https://doi.org/10.1016/j.jdmm.2020.100412

Walia, S., Sharma, D., & Mathur, A. (2021). The impact of service quality on passenger satisfaction and loyalty in the Indian aviation industry. *International Journal of Hospitality and Tourism Systems*, *14*(2), 136–143.

Walter, P. G. (2013). Theorising visitor learning in ecotourism. *Journal of Ecotourism*, *12*(1), 15–32. https://doi.org/10.1080/14724049.2012.742093

Wei, W., Miao, L., & Huang, Z. J. (2013). Customer engagement behaviors and hotel responses. *International Journal of Hospitality Management*, *33*, 316–330. https://doi.org/10.1016/j.ijhm.2012.10.002

Wen, J., Huang, S. S., & Ying, T. (2019). Relationships between Chinese cultural values and tourist motivations: A study of Chinese tourists visiting Israel. *Journal of Destination Marketing & Management*, *14*, 100367. https://doi.org/10.1016/j.jdmm.2019.100367

Wolf, J. (2011). Sustainable supply chain management integration: A qualitative analysis of the German manufacturing industry. *Journal of Business Ethics*, *102*(2), 221–235. https://doi.org/10.1007/s10551-011-0806-0

Woodside, A. G. (2017). *The complexity turn: Cultural, management, and marketing applications*. Springer.

Wu, X., Qiao, S., & Tan, Q. (2022). Destination management for ecotourism activity using analytical hierarchy process. *Scientific Programming*, *2022*, 1–6.

Xu, Y., Zhang, Z., Law, R., & Zhang, Z. (2020). Effects of online reviews and managerial responses from a review manipulation perspective. *Current Issues in Tourism*, *23*(17), 2207–2222. https://doi.org/10.1080/13683500.2019.1626814

Annexure: A set of keywords defining variables

Variable	Keyword
Responsiveness	Prompt, fast, helpful, help, willing, service, time
Reliability	Sincere, timely, effective, promise, interest
Empathy	Attention, attentive, personal, convenience, understanding,
Tangible	Local, influence, entertainment, appealing, visual, attire, food, comfortable, fun, activity
Assurance	Safe, information, knowledge, confident, polite, trust, care, courteous
Eco tangible	Degradation, recycle, reuse, eco-friendly, environment-friendly, safe, environmentally safe
Revisit engagement	Again, visit, go, come back, return
Learning engagement	Learn, knowledge, new thing, information, sharing
Recommendation Engagement	Recommend, good, best, beautiful place, appreciate
Eco activity-based learning	Activity, mountain, ride, bird watching, cycling, guided tour
Eco-consciousness	Curiosity, Curious, Discover, explore, search, involve, natural

Part II

Challenges and critiques in ecotourism

NGOs in ecotourism: patrons of sustainability or neo-colonial agents? Evidence from Africa

Amare Wondirad, Denis Tolkach and Brian King

ABSTRACT

NGOs have become involved in ecotourism because of its potential to balance economic development, environmental conservation, and socio-cultural revitalisation. However, some critics have associated ecotourism with neo-colonialism and with the perpetuation of economic and political hegemonies because the concept has been advanced from the West. The present study adopts a qualitative research approach to explore the merits of two opposing views – that NGOs facilitate and advance sustainable development or that they are agents of neo-colonialism. The researchers focus on a nation which was spared the experience of colonisation – Ethiopia – to explore whether ecotourism practice can be accurately characterised as 'neo-colonial'. Through a close examination of NGO involvement in ecotourism, the authors challenge the widely held view that NGOs use sustainable development as a pretext to promote neo-colonial ideas. The paper contributes to theory and practice by explaining the relationship between neo-colonialism and ecotourism. Implications and opportunities for future research are also discussed.

Introduction

The relationship between ecotourism and neo-colonialism has been an enduring interest amongst tourism researchers (Devine & Ojeda, 2017; Loperena, 2017; Mkono, 2019). Defined as a responsible journey to natural areas that aims to support environmental conservation, improve local well-being and educate tourists (The International Ecotourism Society, 2018), ecotourism has been closely associated with sustainable rural development, particularly in the developing nations (Eshun & Tagoe-Darko, 2015; Mitchell & Ashley, 2010). Researchers have documented the ecotourism features that drive sustainable development, including its capacity to contribute to environmental conservation, while retaining authenticity (Mondino & Beery, 2019; Ruhanen, 2013). Stronza and Gordillo (2008) identified successful ecotourism projects in South America, while Snyman (2014) and Tran and Walter (2014) have documented the benefits that ecotourism has delivered in Southern Africa and Vietnam respectively.

However, ecotourism has also been characterised and critiqued as being the product of a Western 'classical conservationist' paradigm, with its fundamental principles and assumptions rooted in Western cultural, economic and political ideologies. This serves as a challenge to the universal applicability of ecotourism (Devine, 2017; Devine & Ojeda, 2017; Manyara & Jones, 2007). As noted by McKercher (2010), the term ecotourism first appeared publicly in a Parks Canada document in 1978. The agency used the term ecotourism to explain visitor enjoyment of natural resources and to demonstrate the inextricable link between nature and humankind (McKercher, 2010). This implies that the conceptual roots of ecotourism originated in Western conservation philosophy. While ecotourism seeks to achieve sustainable development by unlocking opportunities that generate alternative incomes, and encourage local community empowerment and participation (Bien, 2010; Butcher, 2011; Xu, Cui, Sofield, & Li, 2014), other empirical studies have shown an alternative picture – that ecotourism development has facilitated the implantation of neo-colonial policies through expropriation and privatisation of nature, culture and other community resources (Devine, 2017; Loperena, 2017; Ojeda, 2012). Such practices are consistent with neo-colonialism, as they involve the imposition of externally conceived development policies and the systematic marginalisation of communities from their resource ownership, thereby triggering deprivation and dependency (Janzer & Weinstein, 2014; Nkrumah, 1965).

Non-governmental organisations (NGOs) have been actively involved in ecotourism in various developing countries. Their roles include providing financial and technical support, drafting integrated development plans that include an ecotourism component, raising awareness, lobbying governments, assisting local community with projects and building community capacity (Simpson, 2008a, 2008b; Spenceley, 2008). However, critics have long accused the NGOs that operate in developing countries of being agents of neo-colonialism, especially in the case of those receiving overseas funding (Devine & Ojeda, 2017; Easterly & Pfutze, 2008). The origin of funding is often contentious. In Russia, NGOs that are in receipt of overseas funding are classed as 'foreign agents', leading to a lack of funding which limits their capacity to deliver their mission (Daucé, 2015). From the preceding discussion, it is evident that NGOs may on the one hand drive sustainability by providing capacity building, capital or materials that enable local communities to improve their socio-cultural, economic and environmental resources, or represent neo-colonialism by imposing policies and practices that do not benefit local communities and/or the physical environment.

International development programmes including UNWTO's Sustainable Tourism – Eliminating Poverty (STEP), have been criticised for providing fertile ground for the progression of neo-colonialism under the guise of sustainable development assistance (Devine & Ojeda, 2017; Manyara & Jones, 2007). Other scholars have pointed to the influences of NGOs as facilitators of neo-colonialism as a major reason why ecotourism has failed (Devine, 2017; Loperena, 2017). Despite the strong views expressed on either side of the argument, the links between ecotourism and neocolonialism have not been examined in the context of a never-colonized country. Given this gap, the current study offers a distinct and useful contribution to the literature. The researchers undertake an empirical examination of whether NGOs involved in ecotourism development are patrons of sustainable development or serve as agents of neo-colonialism. By addressing this gap in the literature, the authors are able to provide a holistic understanding of if and how NGOs contribute to sustainable ecotourism in developing countries. The authors draw upon Nkrumah's (1965) theory of neo-colonialism and use a non-colonized developing country as the research setting. Specifically, the researchers aim to:

(1) examine the NGO understandings of ecotourism in developing countries;
(2) investigate the implementation of NGO values and principles through the medium of ecotourism;
(3) examine NGO involvement in ecotourism including their approaches and contributions;
(4) explore impediments to NGO contributions for ecotourism; and
(5) propose means to overcome setbacks and scale-up NGO contributions to sustainable ecotourism in Southern Ethiopia.

Literature review

Neo-colonialism and its attributes

The liberation from formerly dominant European colonisers that followed World War II (WWII) was superseded by a wave of neo-colonialism across the developing world including in Sub-Saharan Africa where countries underwent structural adjustments that advanced neo-colonial agendas (Hanson & Hentz, 1999; Smith, 2010). Neo-colonialism extends previously dismantled colonial hegemonies and entitlements through international organisations such as the International Monetary Fund and World Bank (Hanson & Hentz, 1999; Nega & Schneider, 2014; Wijesinghe, Mura, & Bouchon, 2017). Janzer and Weinstein (2014, p. 338) defined neo-colonialism as an 'influence over a population, community, or society in the absence of direct, obvious or formal control'. Nkrumah (1965, p.xi) characterised neo-colonialism as the worst form of imperialism since 'for those who practice it, it means power without responsibility, and for those who suffer from it, it means exploitation without redress'. According to Wijesinghe et al. (2017), neo-colonialism is a practice where former colonisers or new emerging superpowers subtly impose their interests and enforce economic, political and cultural dominance. Similarly, Lassou and Hopper (2016) understood neo-colonialism as a modern manifestation of colonialism that occurs when colonial powers subtly control the political and economic institutions of former colonies with the intention of creating perpetual dependency. They adopt tactics such as an aid that does not aim to develop but to strengthen the infrastructure required for neo-colonial exploitation. Neo-colonialism allegedly employs a variety of methods such as paying the cost of running the state, infiltrating civil servants who occupy positions where they can dictate policies and imposing monetary control over a foreign exchange (Akama, Maingi, & Camargo, 2011; Antwi-Boateng, 2017; Lassou & Hopper, 2016; Nkrumah, 1965). As noted by Devine and Ojeda (2017), Kline and Slocum (2015) and Ojeda (2012), neo-colonial agendas can be transported via NGOs on the pretext of sustainable development and conservation assistance. International Financial Institutions (IFIs) such as the World Bank (WB) and the International Monetary Fund (IMF) dictate developing nation domestic policies and serve as neo-colonial proxies by offering loans to local policies on condition of adjustment (Hanson & Hentz, 1999; Nega & Schneider,

2014; Wijesinghe et al., 2017). Neo-colonial ideas can be disseminated through policy-related activities such as technical advice, donations, loans and market negotiations (Financial Times, 2019; Hanson & Hentz, 1999).

In the operation of neo-colonial projects, foreign capital is deployed to exploit local economic wealth instead of advancing developing nations (Nkrumah, 1965). Neo-colonial investment projects are destined to increase dependency on overseas governments and organisations, rather than fostering local economic development and self-reliance (Loperena, 2017; Manyara & Jones, 2007; Rudovsky, 2015; Southgate, 2006). Ecotourism development projects in Nepal and Kenya have been critiqued as being neo-colonial because communities have been manipulated and because ecotourism resources have been exploited by external operators, and thereby increasing inequality and dependency (Chan & Bhatta, 2013; Manyara & Jones, 2007; Southgate, 2006). Similarly, Devine (2017), discussed how non-traditional tourism segments such as ecotourism and cultural tourism, set an auspicious stage for the spread of neo-colonial ideologies and reforms that expedite the practices of spatial colonisation, land dispossession, and the commodification of culture and identity as objects of tourist consumption in Guatemala. Loperena (2017) also noted how neo-colonial policies have been implemented under the guise of ecotourism and sustainable tourism development while creating the conditions for extractivism, natural resource expropriation, enclosure, and dispossession resulting in disparity, widespread environmental degradation, and ecological devastation.

Communities in developing countries have been denied access to material subsistence, through items such as land, water, food, timber and game on the basis that their expropriation is needed to conform with sustainable tourism principles (Devine & Ojeda, 2017; Loperena, 2017; Ojeda, 2012; Rudovsky, 2015). With little bargaining power, few choices and a tourism sector that is driven by neo-colonial interests in exploiting resources, locals in developing countries are confined to lower-paying and often physically demanding tourism jobs, while jobs with more career-oriented and meaningful roles are occupied by non-locals (Devine & Ojeda, 2017; Southgate, 2006; Tolkach & King, 2015).

NGOs in developing countries

Various authors have documented the constructive roles of NGOs in addressing socio-economic and environmental problems in developing countries (Nega & Schneider, 2014; Vivian, 1994). Though they cannot replace the state, NGOs, charities and development agencies can play support roles in inducing positive change (Nega & Schneider, 2014; Smith, 2010). Some NGOs have advocated sustainable development practices that improve community livelihoods while conserving cultural heritage and natural habitats (Mukherji, 2018). In the case of Africa, there has been a substantial increase in the number and influence of NGOs since the 1980s (Hearn, 2007; Smith, 2010). The impetus has arisen from a push to reduce public expenditures by international financial institutions such as the World Bank and IMF. This forced indebted African states to open their doors to NGOs (Matthews, 2017). It also led to a mushrooming of nonprofit organisations throughout Africa (Ariti, van Vliet, & Verburg, 2018; Matthews, 2017). In particular, the limited capacity of African governments to serve diverse community needs in remote rural areas or to manage natural ecosystems led to a proliferation of NGO activities (Ariti et al., 2018; Obiyan, 2005). NGO involvement in the delivery of environmental services has been a response to the incapacity of many post-colonial governments (Novelli, 2015).

Views are deeply polarised about the effectiveness of NGO engagements in multi-national corporate investment and bilateral/unilateral aid to developing countries (Burns, 1999; Fortanier & Van Wijk, 2010). Some authors have endorsed an integral role for NGOs as one of the fastest-growing and influential forces for development (e.g. Burns, 1999; Halpenny, 2003; Makoba, 2002; Matthews, 2017; Stone, 2015; Trejos, Huang, & Chiang, 2008; Zhuang, Lassoie, & Wolf, 2011). However, others have questioned their real impacts on development (Burns, 1999; Devine, 2017; Gardner, 2012; Hearn, 2007; Livernash, 1992; Ojeda, 2012; Smith, 2010). According to Burns (1999), NGOs work with and on behalf of the grassroots in developing nations. Similarly, Suresh, Babu, and Siva (1999) have characterised NGOs as being motivated to address contemporary societal challenges by differentiating their approaches from those of large, highly structured institutions such as governments. NGOs may also represent specific local and international interest groups across diverse programmes ranging from poverty alleviation, environmental conservation, food production, capacity building and consulting on policy and planning issues (Burns, 1999; Kennedy & Dornan, 2009; Suresh et al., 1999). Their work can involve tangible short-term local projects or may have a broader and longer-term orientation, for example in nurturing awareness, training, social capital creation, and institutional development (Horochowski & Moisey, 1999; Stone, 2015; Suresh et al., 1999; Zhuang et al., 2011).

NGOs have also attracted criticism for being insufficiently representative or accountable to their intended beneficiaries – the poorest and most marginal (Cook, Wright & Andersson, 2017; Singh, 1999; Trejos et al., 2008). They may focus excessively on shorter-term remedies rather than on dealing with underlying problems thereby compromising sovereignty and development prospects. Scholars have also critiqued the inadequate contributions of NGOs where the opportunities that they generate flow to elite community members and experts from the developed nations at the expense of the intended beneficiaries (Manyara & Jones, 2007; Matthews, 2017; Southgate, 2006). Such criticisms offer support for Nkrumah's (1965) neo-colonial argument. Scholars such as Hearn (2007), Livernash (1992), Manji and O'Coill (2002), Ojeda (2012) and Smith (2010) have claimed that NGOs contribute only marginally to poverty alleviation in developing countries, whilst simultaneously undermining their struggles against economic, social and political oppression and dependency. Furthermore, as suggested by Loperena (2017), Nega and Schneider (2014) and Smith (2010), NGOs may contribute negatively to the socio-economic development of developing countries by helping corrupt dictators to remain in power and by cultivating local corruption. Moreover, many developing country governments frustrate NGO efforts to reach the grassroots and to deliver what is expected, leading to ineffectiveness and susceptibility to failure (Ariti et al., 2018; Clark, 2000; Livernash, 1992; Mitlin, Hickey, & Bebbington, 2007).

NGO engagements in ecotourism

Ecotourism in its current guise emerged as an alternative form of tourism development from around the 1980s. The term refers to a specific nature-based product niche that embraces sustainability principles. These principles may be explained using the triple-bottom-line idea (that embraces environmental, social and economic principles). Ecotourism progressively gained attention globally because it was strongly advocated by many academics, gradual acceptance by consumers and a dawning by industry of the destructive impacts of mass tourism. However, despite gaining greater international recognition, ecotourism is widely misunderstood, partly because there is no commonly used definition and also the lack of any precise measurement of its impacts (Baral, Stern, & Hammett, 2012; Buckley, 2009; Tsaur, Lin, & Lin, 2006). This gap has allowed unethical players to misrepresent the concept and to engage in opportunistic 'green-washing'. Furthermore, it has been difficult to provide an holistic measurement of the sustainability ideal since the economic impacts are typically observable in the shorter-term, while measuring the environmental and socio-cultural impacts requires more longitudinal investigations (Butler, 1999). Despite the challenges, NGOs are amongst the various stakeholders that have continued to advocate on behalf of ecotourism as an appropriate type of tourism.

Tourism-oriented NGOs are increasingly prevalent in developing countries as an alternative and supportive agent of development (Kennedy & Dornan, 2009; Zhuang et al., 2011). For instance, NGOs supported community-based rural tourism in Costa Rica through promotion and marketing, capacity building and technical assistance, and lobbied the political actors to make commitments to the sector (Trejos et al., 2008). Studies in Bangladesh by Hassan and Forhad (2013) and by Roy, Raquib, and Sarker (2017) suggest that NGOs have made a positive contribution to sustainable tourism development. Similarly, Stone (2015) credited NGOs as facilitators of ecotourism stakeholder collaboration in Botswana, which helped to organise and empower local communities and provide financial support.

As noted by Halpenny (2003), NGOs that operate in collaborative working environments and in a supportive political climate can contribute to urgently-needed frontiers of development such as finance and technical advisory services to local authorities, capacity building and ecotourism training. While NGOs, such as the World Wide Fund for Nature (WWF), ActionAid and USAID function at a global level, others operate at regional, national and local scales. In practice, most of the NGOs operating in the developing world are relatively small and community-based (Livernash, 1992; Nega & Milofsky, 2011).

Sustainable development and neo-colonialism are intricately interconnected because various loopholes are prevalent in developing countries that allow NGOs to use sustainable development as a pretext to realise 'hidden' agendas (Devine, 2017; Wijesinghe et al., 2017). In conclusion, whilst there are divisions within the literature about the contribution of NGOs to sustainable development, the current study will undertake an empirical examination of whether NGOs in developing countries are patrons of sustainability or mediums of neo-colonialism.

Research context

Ethiopia is an East African country that has not experienced direct colonial rule. Its recent development of tourism has involved strong NGO participation (Feseha, 2012; Frost & Shanka, 2002; Nega & Milofsky, 2011). It is unique in Africa for having retained indigenous values and remarkable cultural heritage for millennia (Mann, 2006; Mitchell & Coles, 2009). Unique features include its calendar, alphabet, language, cuisines and countless authentic festivals along with several historical, cultural

Table 1. Background of interview participants.

Participant No.	Participant category	Sex	Age	Participant location
1	Local community_L1	F	20–30	Hawassa
2	Local community_L2	M	41–50	Arbaminch
3	Local community_L3	M	50^+	Konso
4	Local community_L4	M	31–40	Hawassa
5	Local community_L5	F	31–40	Arbaminch
6	Government organization (EWCA)_G1	M	41–50	Arbaminch
7	Government organization (MoCT)_G2	M	41–50	Arbaminch
8	Government organization (MoFE)_G3	M	50+	Hawassa
9	Government organization (MoCT)_G4	M	31–40	Konso
10	Government organization (MoCT)_G5	M	31–40	Hawassa
11	Private ecotourism enterprise_P1	M	41–50	Hawassa
12	Private ecotourism enterprise_P2	F	20–30	Hawassa
13	Private ecotourism enterprise_P3	M	31–40	Arbaminch
14	Private ecotourism enterprise_P4	M	31–40	Konso
15	Private ecotourism enterprise_P5	M	31–40	Konso
16	NGO_N1	M	41–50	Arbaminch
17	NGO_N2	M	31–40	Konso
18	NGO_N3	M	50^+	Hawassa
19	NGO_N4	M	41–50	Arbaminch
20	NGO_N5	M	41–50	Konso

Note: Government Organization consisted of participants from the Ministry of Culture and Tourism (MoCT); Ethiopian Wildlife Conservation Authority (EWCA) and Ministry of Forest and Environment (MoFE).

and natural attractions (Feseha, 2012; Tegegne, Moyle, & Becken, 2018).

There is a longstanding involvement of NGOs in Ethiopia dating back to the 1960s with the International Red Cross and Save the Children (Waktola, 1999). However, it was after the devastating 1984–1985 drought which triggered the country's worst-ever famine, that prompted the rapid growth involvement by international NGOs (Clark, 2000; Waktola, 1999). At the time of writing, there are approximately 800 foreign NGOs and 2,800 domestic NGOs registered in Ethiopia (G5, January 2018). NGOs progressively redirected their efforts from relief to development as conditions across the country improved (Waktola, 1999). As these circumstances arose, international NGOs began to support the prospects of sustainable tourism for economic development (WTTC, 2019). The present study has been undertaken in the autonomous region of Southern Ethiopia. The region is bestowed with a range of cultural and natural resources and is a major corridor for ecotourism development (ETO, 2017; Wondirad, 2017). The region is also home to 56 tribes and is known for its diverse and exotic cultural and traditional assets (Wondirad, 2017).

Research method

The current research adopts a qualitative approach to examine the extent to which NGOs use ecotourism as a tool to extend neo-colonial ideologies. The study seeks to establish a deeper understanding of the nexus between ecotourism and neo-colonialism by examining NGO roles, values, and interests from the standpoint of key informants in the context of a non-colonized developing country (Maxwell, 2013). Data were gathered using in-depth interviews of 20 purposively selected key informants (5 participants each from NGOs, local communities, private ecotourism enterprises, and governmental institutions, see Table 1). It is believed that these respondents possess adequate experience and information about issues relating to the research objectives (Bernard & Ryan, 2010). The researchers undertook in-depth interviews with questions addressing issues such as NGO engagements in and contributions to ecotourism, the fundamental values of NGOs, cooperation between NGOs and other ecotourism stakeholders, and challenges facing NGOs when developing ecotourism in Southern Ethiopia. The guidelines for in-depth interviews were generated from the existing literature and were adapted to the current research context over the course of various field visits by the researchers.

Data collection was undertaken from December 2017 through August 2018 and the researchers made multiple site visits, namely Hawassa, Arbaminch, and Konso with the aim of understanding the study areas (Grossoehme, 2014; Lincoln & Guba, 1985; Noble & Smith, 2015). The sites represent the tourism destinations with prime attractions and strategic locations along the Southern Ethiopia tourism route (ETO, 2017). This contrasts with the Northern historic route because of its fame for immense natural and exotic cultural attractions that are suited to ecotourism (Lonely Planet, 2016; Tamene & Wondirad 2019; Young, 2012). Consent was sought and obtained from all study participants. The researchers provided assurance that respondent privacy would be respected, and that any information would be used exclusively for academic purposes.

In-depth interviews were recorded and transcribed for purposes of subsequent analysis. Thematic analysis was applied to develop theoretical frameworks as a form of grounded theory. To ensure the trustworthiness of the findings, the researchers deployed a variety of strategies including pilot testing, member checks, confirmability audits, frequent site visits with extended stays, and thick description (Berg, 2007; Guba, 1990; Kebete & Wondirad, 2019). After organising, cleaning and polishing the relevant inputs, the researchers coded the filtered associated data using QDA Miner qualitative data analysis software, version 5.0.20. Through the process, three types of coding were used (open, axial and selective), in line with the suggestions of Brotherton (2008), Easterby-Smith, Thorpe, and Jackson (2015) and Merriam (2009) with a view to reaching ultimate abstraction and/or theorisation.

Findings and discussions

Rationale for NGO involvement in ecotourism

Ecotourism has recently emerged as a preferred type of tourism development for NGOs operating within Ethiopia. They seek to facilitate local economic development through the sustainable utilisation of natural and cultural resources. In this respect, NGOs see themselves as providers of support to the government and local communities in the development of sustainable ecotourism. Most NGO participants expressed their view that local communities, private ecotourism enterprises, and government institutions have limited understanding of ecotourism. However, NGOs themselves expressed inconsistent understandings of ecotourism. Some considered the consumptive use of natural resources, such as hunting as legitimate:

> We understand ecotourism as a branch of tourism interwoven with nature and biodiversity, which involves hunting and wildlife tourism. We believe that if ecotourism is properly planned, developed and managed with comprehensive stakeholder participation, it can bring both economic and environmental benefits. (N1, May 2018).

Some researchers (e.g. Butler, 1991; Fennell, 2001; Hoenegaard, 1994; Novelli, Barnes, & Humavindu, 2006) have observed that ecotourism may include the consumptive use of resources such as through hunting, in line with conservational policies and practices. The intention is to bring economic benefits that will improve the livelihood of communities. However, Mkono (2019) recently associated hunting and neo-colonialism since hunting schemes grant tourists more power and access to wildlife resources relative to locals. This leads to discontent towards hunting within the wider community since it is perceived as a (neo-colonial) practice that privileges Western elites to exploit native wildlife resources (Mkono, 2019). In this connection, one private sector representative (P5, March 2018) expressed concern that 'communities who reside in the vicinity of national parks where wildlife hunting is practiced, are resentful since they believe that they are denied access to resources while the government permits hunting for tourists because tourists can pay money'. This is consistent with the hostile relationship between protected areas and local communities in Southern Ethiopia that was observed by Yitbarek, Tadie, Timer, and Fischer (2013). As Yitbarek et al. (2013) stated, the government of Ethiopia has benefited from a steady flow of revenues from national parks within the region, without implementing a fair and equitable benefit sharing scheme for the surrounding communities. Local communities, on the other hand, must tolerate costs such as restricted access and prohibitions on hunting and on using the land for agricultural and grazing practices (Yitbarek et al., 2013). Meanwhile, given the lifestyles of most indigenous tribes in the region, the contributions made via financial compensation may be perceived as unattractive, thereby making benefit-sharing mechanisms difficult to implement. The instances where the local community has reduced access to natural resources for tourist benefit has neo-colonial overtones. This is despite the potential attribution of the problem to poor destination management by the respective federal, regional and local governments in Southern Ethiopia.

NGO values and ecotourism development principles

Well defined and established principles and values are critical to guide organisational operations effectively. In light of that, the current study examined what values and principles do NGOs have as a blueprint to plan and guide their mission in a more efficient manner. Given, every consequence is a byproduct of a certain action; crafting and embracing organisational values that carefully consider sustainability principles and promote collaboration, fairness, and equity among stakeholders foster the development of sustainable ecotourism (Faux & Dwyer, 2009; Novelli, 2015). In the case of Southern Ethiopia, NGO respondents stated that despite several constraints that undermine efficiency, they seek to operate in line with sustainable development principles. As core elements of their organisational values, NGOs intend to:

(1) Promote the development of environmentally and socio-culturally responsible and economically competitive ecotourism
(2) Ensure community participation, empowerment and livelihood improvement, and
(3) Promote stakeholder interaction as the following mission statement of an international NGO explains.

> German Society for International Cooperation (GIZ) envisions identifying ecologically threatened areas of Ethiopia and setting up economically productive projects through key stakeholder collaboration including active community participation to provide locals with complementary income to discourage deforestation and other environmentally destructive activities (N4, August 2018).

Guided by such principles, NGOs are contributing to the sustainable development of ecotourism in Southern Ethiopia. Interestingly, 75% of the interviewees viewed NGOs as making a better contribution to sustainable development through ecotourism than local, regional and federal governments. One community representative (L5) stated that the Ethiopian government has not yet made any significant efforts to develop sustainable and competitive ecotourism in the region, apart from collecting income. To substantiate this view, the participant cited the absence of tourist information centres in major ecotourism destinations across Southern Ethiopia, despite ongoing visitor and community demand. Apart from the UNWTO-developed Konso tourist information centre (see Figure 2), (in 2010) (UNWTO, 2017), no tourism information centre has been established even in prime destinations such as Arbaminch and Hawassa.

A notable example of an NGO bottom-up, community-oriented tourism development approach (in contrast with neo-colonial approaches) was the benefit-sharing scheme developed and implemented in Konso by the UNWTO and Ethiopian Sustainable Tourism Alliance (ESTA) in 2006. Though the agreement was ultimately violated by the government in 2012, leading to a series of massive public protests and destruction in Konso (Addis Standard, 2016; Dire Tube, 2016), during its existence it had successfully reconciled competing local community and government institution interests through continuous dialogue and negotiations (Wondirad, 2018). As stated by one local community representative (L3, May 2018), based on this negotiation, 80% of the tourism revenue should be distributed within the local community while the remaining 20% being allocated for district administration.

Given the nascent state of ecotourism in Southern Ethiopia, it is notable that some NGOs embrace the triple-bottom-line concept. Nevertheless, what matters most is whether such values are executed properly and consistently. This should involve working towards the creation of a stakeholder platform in which key actors participate to discuss issues of importance to them and to the ecotourism sector. One local community participant emphasised the following in the face of rapidly declining natural and cultural attractions in the region:

> In my opinion, most of the problems we are facing these days could have been tackled before getting worse. Given most of the challenges of ecotourism development in our area are quite interrelated, a sustained discussion among key stakeholders is very crucial. If some of the actors play a positive role while others are deconstructing, change can never be realized (L4, July 2018).

Halpenny (2003) and Zhuang et al. (2011) noted that the success of NGO initiated projects depends on the existence of an enabling and supporting environment, the mobilisation of adequate resources and openness to work in collaboration with communities, policymakers, authorities and the private sector. Other researchers have identified that it will be worth establishing platforms for continuous discussion between stakeholders to clarify any issues and ultimately to reach common ground (Bramwell, 2011; Weaver, 2006). It has been observed that the best prospects for NGOs to play a constructive role in ecotourism in developing countries is when they are properly monitored and are supported by other stakeholders through ongoing discussions with various levels of government, the private sector, and local communities (Halpenny, 2003; Xu et al., 2014; Zhuang et al., 2011). Successive rounds of discussion amongst stakeholders can help to clear up potential misunderstandings and to build trust for the purposes of further ecotourism collaborations and partnerships. This would potentially address a current challenge facing the development of ecotourism in Southern Ethiopia (Jamal & Getz, 1995; Stone, 2015; Waayers, Lee, & Newsome, 2012)

NGO cooperation with other stakeholders

The findings have noted an absence of effective and consistent NGO cooperation with other stakeholders, particularly governments. The Ministry of Culture and Tourism (MoCT) has been identified by foreign NGO participants as the weakest and less responsible in the country. One foreign NGO participant shared the following feelings, which were widely representative:

> Despite our effort to help and provide technical support, the Ministry of Culture and Tourism is not proactive and not keenly responsible for our calls. Local communities

> tend to be open for help but there are communication and cultural barriers between us. Therefore, there is a missing link between us and the communities. Consequently, we feel that we have not contributed to the sustainable development of ecotourism as much as we want to do (N2, July 2018).

As was explained by one private tour guide (P1, June 2018), it has been challenging for communities in Southern Ethiopia to communicate, work, and cooperate directly with international NGOs, due to low literacy rates within most ethnic tribes. Furthermore, inadequate local cultural knowledge amongst NGOs has led to poor understanding between communities and NGOs, which in turn precludes the development of trust (Mkono, 2019). As noted by Nega and Milofsky (2011), local dialogue, trust, and shared decision-making are essential to carry out effective projects locally. One might also argue that in addition to limiting NGO contributions to sustainable ecotourism development, the vacuum created due to poor governance, lack of capacity and inadequate follow-up might serve as a loophole in which NGOs can easily utilise to instil neo-colonial ideologies onto local communities.

The current NGO interactions with private ecotourism institutions are selective and seasonal. The primary opportunities for communication and cooperation between NGOs and the private sector are capacity building, training, and workshops though these are rarely NGO-organised. As was noted by one private sector participant (P3, April 2018), 'Once such workshops are completed, everything goes back to business as usual and through time those parties fail even to recall major issues discussed.' This suggests that the existing interactions between NGOs and other ecotourism stakeholders in Southern Ethiopia is unsteady and sporadic and lacks a clear roadmap and commitment (Italemahu, 2015; Wondirad, 2017). The lack of effective collaboration is mainly attributable to the absence of a responsible organ to advance initiatives and offer leadership in facilitating such partnership endeavours at the grassroots level. As was implied by various research participants, there is a need for continuous progress using different techniques. The following excerpt from a private sector participant exemplifies a popularly held view.

> Given ecotourism is relatively a recent phenomenon in the region, it did not yet come to the stockholders' attention as it should have. Subsequently, collaborative efforts that aim to ensure sustainable ecotourism development in Southern Ethiopia are limited. The existing stakeholder relationships and collaborations are more informal, seasonal and inconsistent (P2, February 2018).

It is evident that care is needed in situations where collaboration and understanding between NGOs and other stakeholders are lacking. There is a risk that NGO activities could become a tool to promote neo-colonial style development, rather than advancing genuine bottom-up developments by empowering local communities, valuing indigenous knowledge and protecting the natural environment (Janzer & Weinstein, 2014).

Participation and contributions of NGOs in Southern Ethiopian ecotourism development

A range of domestic and international NGOs currently support sustainable ecotourism developments. There are eight international NGOs and one domestic NGO operating in Southern Ethiopia (see Table 2). These NGOs support tourism-related infrastructure and facility development, promote local community empowerment, and provide technical support in areas such as land-use planning, protected area management, and integrated destination marketing and promotion (Barkin & Bouchez, 2002; Jepson, 2005; Novelli, 2015; Simpson, 2008a, 2008b). Moreover, NGOs promote the practice of environmental and ecosystem conservation and advocate stakeholder collaboration to stimulate sustainable ecotourism development in the region.

Figure 1 depicts some of the facilities that have been established in Nechisar National Park through NGO financial and technical support (a 5 km paved entranceway to the national park, and a harbour with visitor facilities). Moreover, a speed boat was donated to the national park management for patrolling purposes. As was stated by a participant from the government institutions (G1, January 2018), the donation of speedboats has been instrumental in preventing rampant illegal fishing in Lake Chamo, part of Nechisar National Park. UNWTO has built Konso tourist information centre and designed a website to promote the cultural landscape and associated attractions in Konso (see Figure 2). The same organisation also trained locals as tour guides and provided financial and technical support to community-based ecotourism projects in Dorze Village, Arbaminch. The NGOs' support in Konso, in particular, brought remarkable outcomes from a sustainable ecotourism development perspective.

Before 2006, international tourist arrivals to Konso were below 2,000 a year (UNWTO, 2017). However, following the inauguration of the STEP project, known as Konso Community Tourism Project, which was implemented by Konso Special District with additional technical support from SNV Ethiopia, international tourist arrivals increased by over 100% in three years (from 1,833 in 2006 to 4,354 in 2009). As is demonstrated in recent tourist arrival records obtained from Konso tourist information centre, visitor arrivals have been

Figure 1. Facilities (solar power source and visitor ticket office) developed by NGOs at Nechisar National Park.

Figure 2. Konso tourist information centre built by UNWTO.

growing rapidly – some 16,596 international tourists visited Konso Cultural Landscape in 2015. Another exemplary NGO achievement in Southern Ethiopia is the community-based ecotourism project outcomes of USAID in partnership with ESTA in Central and Southern Rift Valleys, which created additional income to 40,000 community members whilst preserving the cultural heritage and the natural environment (Brookland, 2013). Moreover, the project conducted community capacity building through training and financial support, marketing, and promotion as well as supported the establishment of a local community association (Brookland, 2013). Furthermore, a project known as STRONGBOW has been working towards capacity building in higher education institutions in Southern Ethiopia, focusing on natural resource management, sustainable ecotourism development and other cross-cutting socio-cultural issues such as gender equality and women empowerment.

Investigations of other developing countries have reached similar outcomes about the role of NGOs in sustainable ecotourism development (Barkin & Bouchez, 2002; Halpenny, 2003; Zhuang et al., 2011). A study by Zhuang et al. (2011) in China also discussed the imperative roles of NGOs in sustainable ecotourism development. The authors observed that NGOs have played a coordination role across key ecotourism stakeholders and have mobilised the resources that are necessary for conservation and development. Similarly, an empirical study in Mexico evidenced that a local NGO has

Table 2. NGOs supporting ecotourism development in Southern Ethiopia.

No.	NGO's name	Areas of contribution
1	German Society for International Cooperation (GIZ)	Financial and technical support (park management plan) as well as capacity building to aid conservation effort, enhance management effectiveness and assisting in marketing protected areas in Southern Ethiopia (e.g. Nechisar National Park and Chebera Churchura National Park).
2	Sustainable Tourism Based on Natural Resource Management with Gender Balance Towards Women (STRONGBOW)	Financial support to research projects and continuous professional development in higher education institutions found in the region (Hawassa University and Arbaminch University)
3	Norwegian Programme for Capacity Development in Higher Education and Research for Development	Financial and technical support to research projects and professional development pertaining to ecotourism and natural resource management in Southern Ethiopian Universities including Hawassa University.
4	United Nations World Tourism Organization (UNWTO)	Financial and technical support towards community-based ecotourism projects (e.g. Dorze village in Arbaminch and Konso)
5	SNV Netherlands Development Organization	Financial and technical support (e.g. Konso Community Tourism Project)
6	Linden Trust for Conservation	Financial and technical support (e.g. Nechisar National Park, Arbaminch)
7	Ethiopian Sustainable Tourism Alliance (ESTA)	Financial and technical support (e.g. community-based ecotourism projects in Central and Southern Rift Valleys)
8	United States Agency for International Development (USAID)	Financial and technical support (e.g. community-based ecotourism projects in Central and Southern Rift Valleys)
9	Konso Development Association (KDA)	Local infrastructural development, capacity building and empowerment of communities and advocates community participation in any development in Konso including tourism/ecotourism plus financial and technical support

Generated from in-depth interviews.

successfully developed ecotourism by facilitating community participation and promoting the interaction of groups from different cultural backgrounds and social classes where competing interests of stakeholders were appropriately reconciled (Barkin & Bouchez, 2002). By examining six international cases, Halpenny (2003) substantiated robust NGO contributions where they played crucial roles in conserving natural and cultural resources by providing financial support and technical expertise to local authorities, conducting research on challenges of resource conservation and providing training to ecotourism stakeholders such as communities and the private sector. It is apparent that where NGOs are properly monitored and are provided with necessary support from stakeholders such as the government, they can be decisive in developing sustainable ecotourism. In this respect, the Government of Ethiopia has enforced a regulation about budget allocations, as is underlined in the following extract from an NGO coordinator.

> The contribution of NGOs towards sustainable ecotourism development in Southern Ethiopia is generally appreciable. Most of the time NGOs conduct projects such as watershed management, protected area management, and land use planning, as well as community awareness campaigns. Moreover, they support infrastructural and facility development efforts in and around protected areas. In consultation with park offices, they also identify the ecotourism resource base, facility requirements and challenges related to illegal settlements, deforestation, and unauthorized fishing to improve the region's protected areas condition and thereby boost the ecotourism sector. However, due to a lack of proper support and facilitation, we failed to fully exploit what NGOs could deliver. There is also a culture of corruption within the NGOs themselves. To rectify this concern, the Ethiopian government has ratified the 70/30 regulation regarding NGOs' budget allocation. Accordingly, 70% of the NGOs' budget should be allocated to their respective projects, while they can use up to 30% of their budget to cover administrative expenses (N4, January 2018).

NGOs have contributed towards the sustainable development of ecotourism in Southern Ethiopia in several ways, namely through: (1) providing financial assistance to support tourism infrastructural and facility development, (2) empowering local communities, (3) increasing public awareness about environmental conservation

Table 3. Strengths (achievements) and weaknesses of NGOs in Southern Ethiopian ecotourism.

Achievements	Weaknesses
• Capacity building, community empowerment, as well as sustainable destination development and marketing • Technical and financial support to community-based ecotourism projects • Facilitation of ecotourism stakeholder discussion forums in a couple of sites in Southern Ethiopia such as Arbaminch and Konso to foster mutual understanding • Technical advice and financial contributions in infrastructural development around protected areas to make national parks more attractive • Absence of any land grab practice or expropriation of community resources on the pretexts of 'greener' development	• Lack of sound local knowledge • Lack of well-planned exit strategies • Internal corruption and misuse of funds • Limitations in scaling-up best practices

Generated from in-depth interviews.

and sustainability, (4) advocating a steadfast ecotourism stakeholder collaboration, and (5) providing technical expertise to assist the formulation of protected area management plans and the marketing and promotion of ecotourism products. However, NGOs have some weaknesses that deserve attention for their future success (see Table 3).

Challenges affecting NGO contributions to ecotourism

NGOs in Southern Ethiopia are currently facing acute challenges in developing sustainable ecotourism. One concerns a restrictive working environment since the government often perceives them as agents of foreign interests (Bekele, Hopkins, & Noble, 2009; Halpenny, 2003; Nega & Milofsky, 2011). International powers such as the European Union and the USA have strategic interests in Ethiopia, a result of the country's geographic location in a politically turbulent region of East Africa adjoining Somalia. The latter country has experienced unremitting civil war and political turmoil over two decades and is a breeding ground for religious extremism, terrorism and regional instability (Dagne, 2002; Samatar, 2007; Shinn, 2004). In this regional context, Ethiopia is considered as one of the West's most dependable regional allies, especially against terrorism. However, the government of Ethiopia has been accused of human rights violations and of being totalitarian. Such allegations have usually been exposed by nongovernmental organisations (Bekele et al., 2009; Nega & Milofsky, 2011). This has prompted the Ethiopian government to be suspicious of NGOs and rarely views them positively (Ariti et al., 2018; Clark, 2000; Nega & Milofsky, 2011).

According to Clark (2000), Bekele et al. (2009) and Dupuy, Ron, and Prakash (2015), the government views NGOs in Ethiopia as political opponents rather than as development partners. The Ethiopian government monitors domestic and foreign NGOs closely. A severe Civil Societies' (CSOs) Law was introduced in 2009 (Bekele et al., 2009; Nega & Milofsky, 2011). Various research participants mentioned the government attitude, for example:

> We have constraints in terms of obtaining the required support from the government, both federal to local. We also do not exactly know our jurisdiction in view of the 2009 Civil Societies (CSOs) law. This law is delicate enough to be interpreted differently by the government, which in turn, might cause undesirable consequences. As a result, our movements and interactions with other ecotourism stakeholders including the private sector and local communities are carefully designed and quite limited (N5, April 2018).

As mentioned by Nega and Milofsky (2011), the 2009 CSOs law was intended to regulate NGO activities and has been denounced as restrictive and as violating basic freedoms of association that are enshrined in the country's constitution and other international human rights obligations. The 2009 CSOs law has been described as the most restrictive of all laws passed by any country in Sub-Saharan Africa (Ariti et al., 2018; Bekele et al., 2009; Dupuy et al., 2015). The law prohibits NGOs from participating in any activity pertaining to human rights, women's rights, children's rights, citizenship rights, conflict resolution and policy and governance issues (Bekele et al., 2009; Dupuy et al., 2015). As a result, NGOs have no scope to engage in policy issues in Ethiopia (Ariti et al., 2018; Nega & Milofsky, 2011). That may explain why NGOs have described themselves as supportive stakeholders of ecotourism development in the current study, rather than as key actors. According to the 2009 CSOs law, even local NGOs must abstain from such duties if they receive more than 10% of their funding from foreign sources (Bekele et al., 2009; Dupuy et al., 2015). Consequently, NGOs currently obtain insufficient support from the Ethiopian government. They are also restricted and censored, which compromises their performance given that ecotourism involves a range of issues that are prohibited by the 2009 CSOs law, including politics, governance and community rights. Moreover, the management of many protected areas, such as Nechisar National Park, raises many concerns since the federal government continues to administer the park using a centralised top-down management style, inconsistent with the core principles of ecotourism. Despite NGO opposition to this management philosophy, research participants implied that they have neither advised nor corrected the government for fear of reprisals.

NGOs have also been impeded by financial instability and poor inter-sectoral linkages. Additional challenges stem from inadequate qualified personnel, weak institutional capacity and a lack of proper takeover and continuation of projects. These result in projects being short-term and unsustainable over the longer term. Anticipated deliverables have also been impeded by government corruption and even by corruption within the NGOs themselves (Smith, 2010; Wheeller, 2005). The problem has been reported by the World Economic Forum's Global Competitiveness Report (2016–2017) authored by Schwab (2016) which identified corruption as the number one challenge of doing business in Ethiopia. Most development aid needs to pass through government officials for approval and is highly exposed to corruption and misuse (Asmare, 2016; Hassan, 2013; Mkono, 2019; Smith, 2010). The following NGO research participant outlined the challenges confronting NGOs as follows:

> Administrative issues (unnecessary bureaucracy), perception problem, lack of well-trained human power with strong local knowledge, low level of community awareness, and lack of administrative support are our chronic challenges. Moreover, challenges such as lack of skilled manpower, image problem (the government does not trust NGOs and as a result, it formulated the 2009 CSOs law, which restricts activities of NGOs), and lack of adequately prepared stakeholder to takeover projects are setbacks that limit our effort to support the sustainable ecotourism development in Southern Ethiopia (N2, April 2018).

NGO efforts in other developing countries have also been compromised by corruption, poor governance and lack of well-studied exit plans (Mkono, 2019; Smith, 2010). Ahebwa, van der Duim, and Sandbrook (2012) revealed that in Uganda the compensation for the costs of environmental conservation to local communities is commonly embezzled by corrupt officials. As noted by Smith (2010), in Nigeria, it is normal to associate NGOs with fraud, deceit, and corruption. Corruption also has a significant impact on wildlife management in Africa. Hunting concessions for wildlife management permits are commonly administered through officers by bribery (Mkono, 2019). Due to the previously mentioned factors, the contributions that NGOs have made towards sustainable ecotourism development in Southern Ethiopia have been significantly constrained. If these setbacks were tackled, more significant positive impacts could have been made on ecosystem conservation, ecotourism product development, social capital creation, community empowerment, and participation, leading to sustainable ecotourism development.

As has been frequently mentioned in the literature, international NGOs that are funded from the developed economies are alleged to prioritise the concerns of their financers, thereby infusing neo-colonial doctrines (Devine & Ojeda, 2017; Dixey, 2008; Easterly & Pfutze, 2008; Halpenny, 2003; Xu et al., 2014). However, when Nkrumah's (1965) theory of neo-colonialism is considered, the current study has not found little concrete evidence in Southern Ethiopia. Instead, NGOs have achieved notable ecotourism development in the region, despite facing massive challenges. As is exemplified by the following statement from a local community representative, NGO efforts are being recognised by many local governments, private ecotourism enterprises, and local communities alike:

> Particularly, NGOs such as the German Society for International Cooperation (GIZ) came to help the ecotourism development in a new approach. They first try to understand the interests of various stakeholders including local fishers, local boat-renting associations, local guide associations, pastoralists, governmental institutions, and local communities. Several discussion platforms were organized by GIZ to understand stakeholders' respective interest before embarking on any kind of ecotourism development (L2, May 2018).

Southern Ethiopia exemplifies a very different picture from what has been presented in the literature where NGOs frequently facilitate the spread of neo-colonialism under the pretext of sustainable ecotourism development. The researchers have documented numerous cases from across the globe in reaching this conclusion (Devine & Ojeda, 2017; Loperena, 2017; Mkono, 2019). This may be attributable to the stringent rules and regulations for NGO adherence. Close monitoring by government means that there is minimal prospect of embedding donor interests. As was widely noted by NGO participants in the study, their efforts to develop ecotourism and to help local communities are sometimes viewed with suspicion. Furthermore, NGOs that support ecotourism in Ethiopia benefit from the guidance and support that is provided by the Ethiopian Wildlife Conservation Authority (EWCA). This may support their effectiveness, given their unfamiliarity with the local context and minimal trust from communities. EWCA is an autonomous public agency of the federal government that is accountable to the Ministry of Culture and Tourism (MoCT) with the jurisdiction of planning, developing, and managing major national parks in the country (EWCA Establishment Proclamation, 2008). NGOs are particularly involved in supporting the planning, development and marketing of national parks in the country, notably in the case of the German Society for International Cooperation (GIZ). On top of the aforementioned causes, the current substantial contribution of NGOs to ecotourism may not accurately reflect their extraordinary work. Rather it may highlight the weaknesses of other sub-regional stakeholders such as the government and private sector. As researchers have recently shown, a lack of government accountability has exacerbated environmental degradation in Southern Ethiopia (Ethiopian Reporter, 2016; Wondirad, 2017). The previous Prime Minister of Ethiopia made the admission by attributing blame to regional political leaders for the loss of unique wildlife resources, which they had pledged to protect and conserve (Ethiopian Reporter, 2016). Ultimately, since the development of ecotourism in Southern Ethiopia is relatively new and NGOs have focused more on providing basic infrastructure and development activities, there are few opportunities to spread foreign ideologies under the guise of ecotourism. Nevertheless, it is equally important to acknowledge that the region's desperate need for socio-economic development and inadequate resources may lead to the acceptance of any development model that functions in the short-run at the expense of longer-term sustainable development. Developing countries often welcome any ecotourism development model for the sake of financial benefits, overlooking longer-term environmental and socio-cultural sustainability (de Haas, 2002; Ruhanen, 2013; Scheyvens, 1999). A government participant in the study reflected on this issue as follows:

> Disagreement exists between park development professionals and politicians regarding which development model to follow. Professionals often advocate the development of infrastructure and facilities that properly support visitors along with ensuring ecosystem and environment rehabilitation followed by marketing to visitors, whereas officials promote the development and marketing of ecotourism resources alongside due to its short-term economic returns (G1, May 2018).

The quotation confirms Mkono's (2019) view that when it comes to spreading neo-colonial ideologies, it is unfair to pin the blame exclusively on Westerners seeking to exploit the economic desperation of Africa. Money-hungry African politicians who lack moral values are equally responsible. The current study findings diverge from the existing literature by revealing that NGOs sometimes do a better job than government and the private sectors in contributing to ecotourism in developing countries. The current study findings challenge some of the previous literature, by providing additional empirical verification to demonstrate a more positive and active contribution.

Conclusion and implications for future research

Though ecotourism has been hailed as a form of tourism that promotes ethical consumption and sustainable practices, it has also been suspect for its neo-colonial connotations (Figueroa & Waitt, 2011). If developing countries are to harness the prospective opportunities of ecotourism, they need to muster a range of support. When such support is sourced from within the developed world, it will usually emanate through NGOs that are attempting to tackle problems such as poverty, unemployment, and environmental conservation (Kennedy & Dornan, 2009). According to Liverman and Vilas (2006) NGO work in development and conservation has generated mixed results. Nevertheless, in the context of southern Ethiopia, the current research has not detected any efforts at spreading neo-colonial agendas or practices featuring the expropriation of land or other resources on the pretext of pursuing 'greener' projects, marking a contrast with previous research (Devine & Ojeda, 2017; Easterly & Pfutze, 2008; Kline & Slocum, 2015; Loperena, 2017; Ojeda, 2012). The current researchers have found NGOs to be supportive of environmental conservation, tourist facility development, and local community empowerment and participation. Relative to the other actors involved in ecotourism in the region, NGOs appeared to be the most supportive of sustainable development. There are several contributing factors. Firstly, NGOs are treated by the Ethiopian government with suspicion and their activities are subject to strict regulation. Secondly, ecotourism in Ethiopia is in its infancy and has grown only modestly. At this stage of development, NGO activities are rather basic and focus on spreading awareness about ecotourism and undertaking infrastructure work to improve the physical environment. Lastly, an Ethiopian organisation, EWCA consults NGOs regarding local context and appropriate activities. The potential contributions of NGOs to the sustainable development of ecotourism in Southern Ethiopia would have been more meaningful if not for the various aforementioned constraints.

Despite the provision of significant NGO contributions to ecotourism development, an urgent strengthening of collaborative working environments is needed, along with the establishment of linkages with local officials, experts, and communities. Since most international NGOs working in Africa have limited local know-how, necessary support includes proper local orientation prior to the start of any new projects (Garland, 2008; Novelli, 2015). In particular, the government should work closely with NGOs during the closing phases of ecotourism projects to ensure a proper handover to local communities. Frequent inspections should also be executed to confirm whether NGO funds are being appropriately allocated and utilised in accordance with the law, and corrective measures need to be taken in cases of noncompliance. NGOs should design and plan their exit strategy well in advance by training and empowering local communities, as well as strengthening local institutional capacities to transfer ecotourism projects (Mitchell & Ashley, 2010; Spenceley, 2008). Furthermore, as pointed out by research participants and discussed in the literature, the absence of uniform metrics to measure the success of NGO efforts in poverty reduction, environmental conservation, and community outreach complicates efforts to ascertain their precise contribution (Kennedy & Dornan, 2009; Suresh et al., 1999). In this regard clearly outlining standards and benchmarks is instrumental to quantify NGO accomplishments and vice versa in concrete terms. Above all, the 2009 preventive CSOs law should be amended to create a conducive working environment in a way that does not compromise national sovereignty and internal security.

While the exact roles and contributions of NGOs in ensuring sustainable development using ecotourism as an instrument are still contestable, the current study suggests that NGOs should focus more on tackling the root causes of the challenges of ecotourism development, instead of depending on short-term 'handouts'. Failure to do so will undermine the ecotourism fundamentals of ensuring long-term economic development and environmental conservation. By exploring the critical roles of NGOs in the sustainable ecotourism development from a unique Research context, the current study has offered substantive theoretical contributions and practical implications to the ongoing discussion of neo-colonialism and ecotourism literature. The researchers have argued that although endemic challenges require domestic solutions that are crafted to fix the underlying causes of socio-economic development problems, foreign NGOs that are properly guided, supported and themed can be also crucial allies in tackling such issues. Future researchers might investigate the origin of community suspicions towards NGOs in developing countries. Moreover, further empirical studies can explore how modern scientific knowledge that is advocated by NGOs can be better integrated with antique indigenous knowledge so that the foundations of ecotourism can be strengthened.

Disclosure statement

No potential conflict of interest was reported by the authors.

ORCID

D. Tolkach http://orcid.org/0000-0002-8820-7764
B. King http://orcid.org/0000-0002-5300-5564

References

Addis Standard. (2016). Commentary – what was troubling Konso? Retrieved from http://addisstandard.com/commentary-what-was-troubling-konso/

Ahebwa, W. M., van der Duim, R., & Sandbrook, C. (2012). Tourism revenue sharing policy at Bwindi Impenetrable National Park, Uganda: A policy arrangements approach. *Journal of Sustainable Tourism*, *20*(3), 377–394.

Akama, J. S., Maingi, S., & Camargo, B. A. (2011). Wildlife conservation, safari tourism and the role of tourism certification in Kenya: A postcolonial critique. *Tourism Recreation Research*, *36*(3), 281–291.

Antwi-Boateng, O. (2017). New world order neo-colonialism: A contextual comparison of contemporary China and European colonization in Africa. *Journal of Pan African Studies*, *10*(2), 177–195.

Ariti, A. T., van Vliet, J., & Verburg, P. H. (2018). What restrains Ethiopian NGOs to participate in the development of policies for natural resource management? *Environmental Science & Policy*, *89*, 292–299.

Asmare, B. A. (2016). Pitfalls of tourism development in Ethiopia: The case of Bahir Dar town and its surroundings. *Korean Social Science Journal*, *43*(1), 15–28.

Baral, N., Stern, M. J., & Hammett, A. L. (2012). Developing a scale for evaluating ecotourism by visitors: A study in the Annapurna Conservation Area, Nepal. *Journal of Sustainable Tourism*, *20*(7), 975–989.

Barkin, D., & Bouchez, C. P. (2002). NGO–community collaboration for ecotourism: A strategy for sustainable regional development. *Current Issues in Tourism*, *5*(3–4), 245–253.

Bekele, Y., Hopkins, C., & Noble, L. (2009). Sounding the Horn: Ethiopia's Civil Society Law threatens human rights defenders. Retrieved from https://www.law.northwestern.edu/legalclinic/humanrights/documents/EthiopiaCSOPaper-Nov2009.pdf

Berg, B. L. (2007). *Qualitative methods for the social sciences* (6th ed.). Boston, MA: Pearson Education, Inc.

Bernard, H. R., & Ryan, G. W. (2010). *Analyzing qualitative data: Systematic approaches*. Thousand Oaks, CA: SAGE Publication, Inc.

Bien, A. (2010). Bridging the gap between policies and the field in sustainable tourism through new international networks. Retrieved from https://www.youtube.com/watch?v=de55JjVZVS8

Bramwell, B. (2011). Governance, the state and sustainable tourism: A political economy approach. *Journal of Sustainable Tourism*, *19*(4–5), 459–477.

Brookland, J. (2013). Community ecotourism destinations open in Ethiopia's Central and Southern Rift Valleys. Retrieved from http://www.counterpart.org/community-ecotourism-destinations-open-in-ethiopias-central-and-southern-ri/

Brotherton, B. (2008). *Researching hospitality and tourism: A student guide*. London, UK: Sage.

Buckley, R. (2009). Evaluating the net effects of ecotourism on the environment: A framework, first assessment and future research. *Journal of Sustainable Tourism*, *17*(6), 643–672.

Burns, P. (1999). Tourism NGOs. *Tourism Recreation Research*, *24*(2), 3–6.

Butcher, J. (2011). Can ecotourism contribute to tackling poverty? The importance of 'symbiosis'. *Current Issues in Tourism*, *14*(3), 295–307.

Butler, R. (1991). Tourism, environment, and sustainable development. *Environmental Conservation*, *18*(3), 201–209.

Chan, R., & Bhatta, K. (2013). Ecotourism planning and sustainable community development: Theoretical Perspectives for Nepal. *South Asian Journal of Tourism and Heritage*, *6*(1), 69–96.

Clark, J. (2000). *Civil society, NGOs, and development in Ethiopia. A snapshot view*. Washington, D.C: NGO and Civil Society Unit Social Development Department, the World Bank.

Cook, N. J., Wright, G. D., & Andersson, K. P. (2017). Local politics of forest governance: Why NGO support can reduce local government responsiveness. *World Development*, *92*, 203–214.

Dagne, T. S. (2002). Africa and the war on terrorism: The case of Somalia. *Mediterranean Quarterly*, *13*(4), 62–73.

Daucé, F. (2015). The duality of coercion in Russia: cracking down on "foreign agents". *Demokratizatsiya: The Journal of Post-Soviet Democratization*, *23*(1), 57–75.

Devine, J. A. (2017). Colonizing space and commodifying place: Tourism's violent geographies. *Journal of Sustainable Tourism*, *25*(5), 634–650.

Devine, J., & Ojeda, D. (2017). Violence and dispossession in tourism development: A critical geographical approach. *Journal of Sustainable Tourism*, *25*(5), 605–617.

de Haas, H. C. (2002). Sustainability of small-scale ecotourism: The case of Niue, South Pacific. *Current Issues in Tourism*, *5* (3-4), 319–337.

Dire Tube. (2016). Ethiopia - Unrest in Konso Region lead the local heritage to ashes. Retrieved from https://www.diretube.com/ethiopia-unrest-in-konso-region-lead-the-local-heritage-to-ashes-sep-15-2016_d13818461.html

Dixey, L. M. (2008). The Unsustainability of community tourism donor projects: Lessons from Zambia. In A. Spenceley (Ed.), *Responsible tourism critical issues for conservation and development* (pp. 325–343). London: Earthscan, UK.

Dupuy, K. E., Ron, J., & Prakash, A. (2015). Who survived? Ethiopia's regulatory crackdown on foreign-funded NGOs. *Review of International Political Economy*, *22*(2), 419–456.

Easterly, W., & Pfutze, T. (2008). Where does the money go? Best and worst practices in foreign aid. *Journal of Economic Perspectives*, *22*(2), 29–52.

Easterby-Smith, M., Thorpe, R., & Jackson, P. R. (2015). *Management and business research* (5th ed.). London: Sage Publications.

Eshun, G., & Tagoe-Darko, E. (2015). Ecotourism development in Ghana: A postcolonial analysis. *Development Southern Africa*, *32*(3), 392–406.

Ethiopian Reporter. (2016). PM Hailemariam Dessalegne blamed regional and local authorities for the destruction of national parks in the country (translation). Retrieved from https://www.ethiopianreporter.com/index.php/content/E18CA0E18985E1888BE18BAD

Ethiopian Tourism Organization. (2017). *Great Ethiopian Routes*. Retrieved from http://www.ethiopia.travel/destinations/ethiopian-routes

EWCA Establishment Proclamation. (2008). A proclamation to provide for the establishment of the Ethiopian wildlife development and conservation authority. Retrieved from https://chilot.me/wp-content/uploads/2011/01/proc-575-ethiopian-wildlife-develop.pdf

Faux, J., & Dwyer, L. (2009). Triple bottom line reporting of tourism organizations to support sustainable development. *BEST Education Network Module*, 1–47.

Fennell, D. A. (2001). A content analysis of ecotourism definitions. *Current Issues in Tourism*, *4*(5), 403–421.

Feseha, M. (2012). *Fundamentals of community-based ecotourism development in Ethiopia*. Addis Ababa: Eclipse Printing and Graphics.

Figueroa, R. M., & Waitt, G. (2011). The moral terrains of ecotourism and the ethics of consumption. *Ethical Consumption*, *260*(274), 260–274. ROUTLEDGE in association with GSE Research.

Financial Times. (2019). *Neo-liberalism is neo-colonialism in disguise*. Retrieved from http://www.sundaytimes.lk/051023/ft/9.html

Fortanier, F., & Van Wijk, J. (2010). Sustainable tourism industry development in sub-Saharan Africa: Consequences of foreign hotels for local employment. *International Business Review*, *19*(2), 191–205.

Frost, F. A., & Shanka, T. (2002). Regionalism in tourism-the case for Kenya and Ethiopia. *Journal of Travel & Tourism Marketing*, *11*(1), 35–58.

Gardner, B. (2012). Tourism and the politics of the global land grab in Tanzania: Markets, appropriation and recognition. *Journal of Peasant Studies*, *39*(2), 377–402.

Garland, E. (2008). The elephant in the room: Confronting the colonial character of wildlife conservation in Africa. *African Studies Review*, *51*(03), 51–74.

Grossoehme, D. H. (2014). Overview of qualitative research. *Journal of Health Care Chaplaincy*, *20*(3), 109–122.

Guba, E. G. (1990). The alternative paradigm dialog. In E. G. Guba (Ed.), *The paradigm dialog* (pp. 17–27). Newbury Park, CA: Sage.

Halpenny, E. A. (2003). NGOs as conservation agents: Achieving conservation through Marine ecotourism. In B. Garrod & J. C. Wilson (Eds.), *Marine ecotourism: Issues and experiences* (pp. 107–120). Clevedon: Channel View Publication.

Hanson, M., & Hentz, J. J. (1999). Neocolonialism and neoliberalism in South Africa and Zambia. *Political Science Quarterly*, *114*(3), 479–502.

Hassan, A., & Forhad, A. (2013). The role of NGOs in the sustainable development in Bangladesh. *Present Environment & Sustainable Development*, *7*(2), 59–72.

Hassan, S. Y. (2013). Aid, Predation and State Capture: The Role of Development Aid in Fueling Corruption and Undermining Governance: The Case of Ethiopia. *Available at SSRN 2303716*.

Hearn, J. (2007). African NGOs: The new compradors? *Development and Change*, *38*(6), 1095–1110.

Hoenegaard, G. (1994). Ecotourism: A status report and conceptual framework. *Journal of Tourism Studies*, *5*(2), 24–35.

Horochowski, K., & Moisey, R. N. (1999). The role of environmental NGOs in sustainable tourism development: A case study in northern Honduras. *Tourism Recreation Research*, *24*(2), 19–30.

Italemahu, T. Z. (2015). *Interaction, institutions and impacts of tourism on the Bishoftu-Modjo-Hawassa route, Central Ethiopia* (Doctoral Dissertation). Addis Ababa University, Addis Ababa). Retrieved from http://etd.aau.edu.et/bitstream/handle/123456789/978/Tesfaye%20Zeleke%20Italemahu.pdf?sequence=1&isAllowed=y

Jamal, T. B., & Getz, D. (1995). Collaboration theory and community tourism planning. *Annals of Tourism Research, 22*(1), 186–204.

Janzer, C. L., & Weinstein, L. S. (2014). Social design and neocolonialism. *Design and Culture, 6*(3), 327–343.

Jepson, P. (2005). Governance and accountability of environmental NGOs. *Environmental Science and Policy, 8*, 515–524.

Kebete, Y., & Wondirad, A. (2019). Visitor management and sustainable destination management nexus in Zegie Peninsula. *Northern Ethiopia. Journal of Destination Marketing & Management, 13*(C), 83–98.

Kennedy, K., & Dornan, D. A. (2009). An overview: Tourism nongovernmental organizations and poverty reduction in developing countries. *Asia Pacific Journal of Tourism Research, 14* (2), 183–200.

Kline, C. S., & Slocum, S. L. (2015). Neoliberalism in ecotourism? *The new Development Paradigm of Multinational Projects in Africa. Journal of Ecotourism, 14*(2–3), 99–112.

Lassou, P. J. C., & Hopper, T. (2016). Government accounting reform in an ex-French African colony: The political economy of neocolonialism. *Critical Perspectives on Accounting, 36*, 39–57.

Lincoln, Y. S., & Guba, E. G. (1985). *Naturalistic inquiry*. Newbury Park, CA: Sage.

Liverman, D. M., & Vilas, S. (2006). Neoliberalism and the environment in Latin America. *Annual Review of Environment and Resources, 31*, 327–363.

Livernash, R. (1992). The growing influence of NGOs in the developing world. *Environment: Science and Policy for Sustainable Development, 34*(5), 12–43.

Lonely Planet. (2016). Lonely Planet names the top 10 countries to visit in 2017. Retrieved from http://www.elledecor.com/life-culture/travel/news/g3344/lonely-planet-names-the-top-10-countries-to-visit-in-2017/

Loperena, C. A. (2017). Honduras is open for business: Extractivist tourism as sustainable development in the wake of disaster? *Journal of Sustainable Tourism, 25*(5), 618–633.

Makoba, J. W. (2002). Nongovernmental organizations (NGOs) and third world development: An alternative approach to development. *Journal of Global South Studies, 19*(1), 53.

Manji, F., & O'Coill, C. (2002). The missionary position: NGOs and development in Africa. *International Affairs, 78*(3), 567–584.

Mann, S. (2006). *Ethiopia, in Makeda's footsteps: Towards a strategy for pro-poor tourism development*. Addis Ababa: World Bank report to Ethiopian Government.

Manyara, G., & Jones, E. (2007). Community-based tourism enterprises development in Kenya: An exploration of their potential as avenues of poverty reduction. *Journal of Sustainable Tourism, 15*(6), 628–644.

Matthews, S. (2017). The role of NGOs in Africa: are they a force for good?. Retrieved from https://theconversation.com/the-role-of-ngos-in-africa-are-they-a-force-for-good-76227

Maxwell, J. A. (2013). *Qualitative research design: An interactive approach* (3rd ed.). Thousand Oaks: SAGE Publication, Inc.

McKercher, B. (2010). Academia and the evolution of ecotourism. *Tourism Recreation Research, 35*(1), 15–26.

Merriam, S. (2009). *Qualitative research: A guide to design and implementation*. San Francisco, CA: Jossey-Bass.

Mitchell, J., & Ashley, C. (2010). *Value chain analysis and poverty reduction at scale: Evidence from tourism is shifting mind-sets*. London: Overseas development institute.

Mitchell, J., & Coles, C. (2009). *Enhancing private sector and community engagement in tourism services in Ethiopia*. London, UK: Overseas development institute.

Mitlin, D., Hickey, S., & Bebbington, A. (2007). Reclaiming development? NGOs and the challenge of alternatives. *World Development, 35*(10), 1699–1720.

Mkono, M. (2019). Neo-colonialism and greed: Africans' views on trophy hunting in social media. *Journal of Sustainable Tourism, 27*(5), 689–704.

Mondino, E., & Beery, T. (2019). Ecotourism as a learning tool for sustainable development. The case of Monviso Transboundary Biosphere Reserve, Italy. *Journal of Ecotourism, 18*(2), 107–121.

Mukherji, R. (2018). Tourism and development: The role of NGOs and the private sector. Retrieved from http://csr-asia.com/newsletter-tourism-and-development-the-role-of-ngos-and-the-private-sector

Nega, B., & Milofsky, C. (2011). Ethiopia's anti-NGO law and its consequences for economic development. *Community Development Journal, 46*(suppl_2), ii33–ii48.

Nega, B., & Schneider, G. (2014). NGOs, the state, and development in Africa. *Review of Social Economy, 72*(4), 485–503.

Nkrumah, K. (1965). *Neo-colonialism: The last stage of imperialism*. New York: International Publishers.

Noble, H., & Smith, J. (2015). Issues of validity and reliability in qualitative research. *Evidence-Based Nursing, 18*(2), 34–35.

Novelli, M. (2015). *Tourism and development in Sub-Saharan Africa: Current issues and local realities*. London: Routledge.

Novelli, M., Barnes, J. I., & Humavindu, M. (2006). The other side of the ecotourism coin: Consumptive tourism in Southern Africa. *Journal of Ecotourism, 5*(1–2), 62–79.

Obiyan, A. S. (2005). A critical examination of the state versus non-governmental organizations (NGOs) in the policy sphere in the global south: Will the state die as the NGOs thrive in Sub-Saharan Africa and Asia? *African and Asian Studies, 4*(3), 301–326.

Ojeda, D. (2012). Green pretexts: Ecotourism, neoliberal conservation and land grabbing in Tayrona National Natural Park. *Colombia. Journal of Peasant Studies, 39*(2), 357–375.

Roy, I., Raquib, T., & Sarker, A. (2017). Contribution of NGOs for socio-economic development in Bangladesh. *Science Journal of Business and Management, 5*(1), 1–8.

Rudovsky, J. F. (2015). The Ecotourism industry is saving Tanzania's animals and threatening its indigenous people. Retrieved from http://www.vice.com/read/casualties-of-conservation-0000649-v22n5

Ruhanen, L. (2013). Local government: Facilitator or inhibitor of sustainable tourism development? *Journal of Sustainable Tourism, 21*(1), 80–98.

Samatar, A. I. (2007). Ethiopian invasion of Somalia, US warlordism & AU shame. *Review of African Political Economy, 34*(111), 155–165.

Scheyvens, R. (1999). Ecotourism and the empowerment of local communities. *Tourism Management, 20*(2), 245–249.

Schwab, K. (2016). *World economic forum global competitiveness report (2016–2017)*. Geneva: World Economic Forum.

Shinn, D. (2004). Fighting terrorism in East Africa and the horn. *Foreign Service Journal, 81*(9), 40.

Simpson, M. C. (2008a). Community benefit tourism initiatives-a conceptual oxymoron? *Tourism Management, 29*(1), 1–18.

Simpson, M. C. (2008b). The impacts of tourism initiatives on rural livelihoods and poverty reduction in South Africa: Mathenjwa and Mqobela. In S. Anna (Ed.), *Responsible tourism critical issues for conservation and development* (pp. 239–266). Trowbridge, UK: Cromwell Press.

Singh, S. (1999). On tourism in Goa—NGOs can make a difference. *Tourism Recreation Research, 24*(2), 92–94. doi:10.1080/02508281.1999.11014882

Smith, D. J. (2010). Corruption, NGOs, and development in Nigeria. *Third World Quarterly, 31*(2), 243–258.

Snyman, S. (2014). The impact of ecotourism employment on rural household incomes and social welfare in six Southern African countries. *Tourism and Hospitality Research, 14*(1–2), 37–52.

Southgate, R. J. (2006). Ecotourism in Kenya: The vulnerability of communities. *Journal of Ecotourism, 5*(1–2), 80–96.

Spenceley, A. (2008). Introduction: Responsible tourism in Southern Africa. In A. Spenceley (Ed.), *Responsible tourism critical issues for conservation and development* (pp. 1–26). London: Earthscan, UK.

Stone, M. T. (2015). Community-based ecotourism: A collaborative partnerships perspective. *Journal of Ecotourism, 14*(2–3), 166–184.

Stronza, A., & Gordillo, J. (2008). Community views of ecotourism. *Annals of Tourism Research, 35*(2), 448–468.

Suresh, K. T., Babu, H., & Siva, S. (1999). Reflections on the role of NGOs in tourism with reference to India. *Tourism Recreation Research, 24*(2), 88–91.

Tamene, K., & Wondirad, A. (2019). Economic impact of tourism on small-scale tourism enterprises in Hawassa City, Southern Ethiopia. *International Journal of Tourism Sciences, 19*(1), 1–24.

Tegegne, W. A., Moyle, B. D., & Becken, S. (2018). A qualitative system dynamics approach to understanding destination image. *Journal of Destination Marketing & Management, 8*, 14–22.

The International Ecotourism Society. (2018). What is ecotourism? Retrieved from http://www.ecotourism.org/what-is-ecotourism

Tolkach, D., & King, B. (2015). Strengthening community-based tourism in a new resource-based island nation: Why and how? *Tourism Management, 48*, 386–398.

Tran, L., & Walter, P. (2014). Ecotourism, gender and development in northern Vietnam. *Annals of Tourism Research, 44*, 116–130.

Trejos, B., Huang, W., & Chiang, L. N. (2008). The role of NGOs in the development of community-based rural tourism in Costa Rica. *Rural Tourism Research, 2*(1), 63–76.

Tsaur, S. H., Lin, Y. C., & Lin, J. H. (2006). Evaluating ecotourism sustainability from the integrated perspective of resource, community and tourism. *Tourism Management, 27*(4), 640–653.

UNWTO. (2017). Konso community tourism project. Retrieved from http://step.unwto.org/ar/node/33474

Vivian, J. (1994). NGOs and sustainable development in Zimbabwe: No magic bullets. *Development and Change, 25* (1), 167–193.

Waayers, D., Lee, D., & Newsome, D. (2012). Exploring the nature of stakeholder collaboration: A case study of marine turtle tourism in the Ningaloo region, Western Australia. *Current Issues in Tourism, 15*(7), 673–692.

Waktola, A. (1999). Exploratory study of two regions in Ethiopia to identify target areas and partners for intervention, Report No. 6, Addis Ababa.

Weaver, D. (2006). *Sustainable tourism*. Amsterdam: Elsevier.

Wheeller, B. (2005). Ecotourism/egotourism and development. In M. Hall & S. Boyd (Eds.), *Nature-based tourism in Peripheral Areas: Development or disaster?* (pp. 263–272). Clevedon: Channel View Publications.

Wijesinghe, S. N., Mura, P., & Bouchon, F. (2017). Tourism knowledge and neocolonialism–a systematic critical review of the literature. *Current Issues in Tourism, 22*(11), 1263–1279.

Wondirad, A. (2017). Who benefits from the ecotourism sector in Southern Ethiopia? *International Journal of Tourism Sciences, 17*(4), 276–297.

Wondirad, A. N. (2018). *Stakeholder collaboration for sustainable ecotourism development in Southern Ethiopia* (Doctoral dissertation). The Hong Kong Polytechnic University, Hong Kong.

World Travel & Tourism Council (WTTC). (2019). Travel & Tourism: Economic Impact. Retrieved from https://www.wttc.org/-/media/files/reports/economic-impact-research/regions-2019/world2019.pdf

Xu, H., Cui, Q., Sofield, T., & Li, F. M. S. (2014). Attaining harmony: Understanding the relationship between ecotourism and protected areas in China. *Journal of Sustainable Tourism, 22* (8), 1131–1150.

Yitbarek, T. W., Tadie, D., Timer, G., & Fischer, A. (2013). Evaluating governance processes in the sharing of revenues from wildlife tourism and hunting in Ethiopia. *Environmental Conservation, 40*(3), 253–265.

Young, J. (2012). Ethiopian protected areas: A "Snapshot". A reference guide for future strategic planning and project funding, 46. Retrieved from https://phe-ethiopia.org/admin/uploads/attachment-1167-Eth%20Protected%20Areas%20Snapshot%201_4_2012.pdf

Zhuang, H., Lassoie, J. P., & Wolf, S. A. (2011). Ecotourism development in China: Prospects for expanded roles for non-governmental organisations. *Journal of Ecotourism, 10*(1), 46–63.

Limits and paradoxes in the development of ecotourism in Kenya: implications on the sustainability of the natural environment

Bonface Odiara Kihima

ABSTRACT

Ecotourism entails travel to natural and relatively undisturbed areas. The government plays a crucial role in delineation and management of such areas. In addition, the private sector sees to it that such areas exist for posterity. In Kenya, one such institution is Ecotourism Kenya (EK) whose motto is to 'link communities, tourism and conservation'. First, this paper takes a critical look at the conservation agenda advanced by EK through the national eco-rating scheme launched in 2002 that seeks to give every tourist accommodation facility a chance to participate, learn and apply best practices of sustainable tourism. Moreover, the theme of community empowerment is discussed based on the accommodation facilities initiated by EK in Kakamega Forest Reserve (240 km^2) in Western Kenya. This paper concludes that, while certification can be a beneficial tool if properly implemented, this is not the case given that there were only 100 establishments certified in Kenya under this scheme by the year 2012. In addition, ecotourism in Kenya needs to dialogue and clearly define its agenda in order to minimize contradictions and dilemmas inherent in ecotourism practice.

Introduction

In today's world, firms are judged based on their business ethics, social accountability and socio-economic awareness as well as financial outcomes (Lee, Hsu, Han, & Kim, 2010). For example, the concept of marketing has evolved from simple selling to societal marketing, driving organizations to endeavour to become responsible and accountable to society (Fyall & Garrod, 2005). Natural areas are no exception since the modern day tourist is said to be environmentally conscious, seeking more for and from green establishments and destinations (Honey, 2008). With this in mind, 'green' accommodation images have become a powerful operational tool in attracting and retaining more guests as well as achieving cost reduction measures through the use of renewable energy and water. Consequently, an eco-rating scheme was introduced in Kenya in 2002, with the view to promoting sustainability of natural areas.

Previously, in the 1990s, in the quest to be more accountable to society, the Kenyan tourism industry established Community Based Tourism Initiatives (CBTIs) whose aim was to enhance the conservation agenda (Manyara & Jones, 2007). Thus, five decades after the first national park was gazetted in Kenya in 1946, the Kenya Wildlife Service (KWS) management came up with the concept of 'parks beyond parks' whose theme was to develop partnership with the local communities in wildlife management (Honey, 2008). This resulted in the formal creation of the first community owned wildlife sanctuary in Kenya, Kimana Group Ranch (Wishitemi & Okello, 2003), which re-affirmed that people and parks can no longer be separated. Ironically, the tourism programme of Kimana conservancy has since disintegrated due to lack of benefits to members (Meguro & Inoue, 2011; Ondicho, 2012). Moreover, nothing much seems to have changed given the poverty levels around protected areas. For example, according to World Research Institute et al. (2007) on average 50% of the people living within a 25 km radius of the most visited parks in Kenya are poor. KIPPRA – Kenya Institute of Public Policy and Research Analysis (2009) on their part admit that community benefits from Masai Mara and Amboseli conservation areas (most visited protected areas in Kenya) are less than 1%.

With this in mind, Cater (1994) and Wheeller (1997) rightfully argued that issues raised by critics of ecotourism remain unanswered and that the basic problems and dilemma witnessed in ecotourism and conservation areas are very much still present. This paper echoes the same sentiments, first by interrogating in depth the eco-rating initiatives in Kenya, and second the general concept of ecotourism development at destination levels. What are the gaps inherent in such initiatives?

Methods

This work is a culmination of personal observations, experiences and encounters as a researcher, scholar and Ecotourism Kenya member. It combines theory and examples from the field through interactions with site managers and Ecotourism Kenya officials between 2012 and 2014. The study adopts an *etic* approach where the researcher as an observer and outsider describes a certain phenomenon (Pearce, 2005), which in this case is ecotourism in the Kenyan context. The paper therefore has limitations in the sense that the researcher did not obtain an *emic* perspective of the eco-tourists, the local communities as well as destination managers in a structured manner.

Consequently, this research largely involved the use of unstructured and spontaneous discussions as well as analysis of the tourism literature and websites. This being a phenomenological approach, it involved individuals who have experienced, lived and interacted with the ecotourism sector. Such an approach enabled the researcher to get details of the topic under study. It included meeting and discussing with various stakeholders in different forums, including the Kenya Coast Tourism Association (KCTA) meetings, Kenya Association of Hotel Keepers and Caterers (KAHC) symposiums, Ecotourism Kenya (EK) workshops and academic seminars touching on the ecotourism subject. As indicated by Mayaka and Prasad (2012), such an approach provides an important insight into the ongoing discourse in a non-intrusive, unrestricted environment. These authors note that 'such unsolicited views can be relied upon to make substantive contribution to a subject matter' (p. 1).

This allowed the researcher to engage with the social world and to view it from the inside-out in order to clearly understand it (Gray, 2004). The researcher therefore engaged with the respondents and probed in detail in order to generate insights into their perspectives of the current ecotourism practices in the country. The method included taking notes during such forums and discussions, comparing and contrasting them to get a common theme. In case more clarification was needed, the researcher engaged the respondents again through face-to-face encounters as well as telephone conversations to clarify on unclear issues.

Arguably, qualitative research has gained prominence as a mode of inquiry into social issues (Ospina & Parry, 2004). Such is grounded in the world of experience where conclusions are derived from other means than statistical analysis with the aim of understanding the social world. This methodology tries to approach reality without preconceived ideas and pre-structured models and patterns, and aims to understand people and not to measure them (Sarantakos, 2013).

The Kenyan destination

Kenya is an East African country located between 4°40′ north and 4°20′ south, and between 34°0′ west and 42°0′ east. It covers a surface area of 582,648 km^2 and is situated astride the Equator. Kenya, whose population stands at 39 million inhabitants (GoK, 2014), is well known for nature-based tourism, otherwise known as safaris (World Resource Institute et al., 2007). It has a littoral frontage of 530 km stretching from Somalia to the Tanzanian border.

Kenya is well endowed with the national park system occupying 44,562 km^2 or 8% of its territory as well as numerous sites of cultural, historical and natural importance. Approximately 70% of visitors come to Kenya to enjoy the country's natural beauty and engage in nature-based activities (World Resource Institute et al., 2007). It is described as the *mzee*, or elder statesman, of nature tourism and ecotourism in Africa (Honey, 2008) as well as the jewel of East Africa (World Resource Institute et al., 2007). In 2013, Kenya received 1.5 million international arrivals, each staying an average of 13.2 days (GoK, 2014) against the projected 3 million international tourists by the year 2012 (GoK, 2007).

Moreover, 80% of inbound tourists to Kenya buy packaged tours (Odunga, 2005), an arrangement whereby they pay overseas tour operators for a complete travel package. Consequently, Ikiara and Okech (cited in World Resource Institute et al., 2007, p. 82) observed that 'most overseas tourists spend a night in Nairobi on arrival, embark on a two-day or so safari to view wildlife, and devote the rest of their holiday to a longer stay on the coastal beaches'. Consequently, beaches and coastal ecosystems continuously account for a large share of tourism earnings, including more than half of all bed-nights spent by tourists in hotel accommodation in 2005 (World Resource Institute et al., 2007). In 2013, most visitors preferred to stay at the coastal region, which accounted for 45.5% of the total bed-nights occupancy, while Nairobi accounted for 24.7% (GoK, 2014).

Linking tourism to conservation

Honey (2008) acknowledged that Kenya has been a trailblazer in ecotourism. This is evidenced by the fact that the first director of KWS – an institution overseeing the management of all national parks in Kenya – was the first head of The International Ecotourism Society (TIES) in the USA. Furthermore, Ecotourism Kenya (formerly, Ecotourism Society of Kenya) was the first non-governmental ecotourism body to be formed in Africa.

The private sector's role, exemplified by Ecotourism Kenya (EK), in driving competitiveness in the tourism sector with regard to green operations is evident. Over the past five decades, certain protected areas in Kenya have been overexploited without due regard to ecological processes. A case in point is the Masai Mara ecosystem in which 108 tourist establishments with a total of 4000 tourist beds have been built (Hoteliers' blatant disregard of law puts Mara ecosystem at risk, 2010). However, it has been noted that the majority (71%) of these establishments function illegally without official authorization. Some 92 establishments (85%) did not carry out prior environmental audits, while 98 establishments (91%) function without a physical plan (Hoteliers' blatant disregard of law puts Mara ecosystem at risk, 2010).

This haphazard development transforms natural environments into victims of tourism development. Consequently, the target to sustainable ecotourism becomes elusive (Cater, 1994) mainly because the environment is still externalized and treated as free goods (Fennell, 2006). It was for these reasons that EK was established in 1996 as a follow-up to the first ever national ecotourism workshop held in Nakuru, Kenya, in 1992. The aim was to form a lobby group that would promote good practices in the tourism industry. This was necessary since, as urged by Plog (1994; and cited in Boniface, 2001, p. 6), 'to protect the world for tourism, and protect the world from tourism, a common set of goals, evolving into specific plans for different regions, is desperately needed'.

The eco-rating scheme

One such goal undertaken by EK was the setting up of the eco-rating scheme. This is a certification process whereby third party assessment is undertaken and written assurance is given that the product, process, service or management system conforms to the standard (Sallows & Font, 2004). Such programmes are characterized by certification fees (Table 1), a logo or a brand (Figure 1), voluntary enrolment, auditing or other forms of assessment (Table 2), and compliance to the regulations that are built into the programme (Fennell, 2006; Sallows & Font, 2004). This institutional dimension helps certified organizations to maintain their legitimacy while conforming to established standards (see also Muriithi, 2014). Initially, this process targeted the accommodation sector in Kenya, after which it was replicated in other countries within the East African region as from 2008.

According to EK, the aim of the scheme is threefold. First, it is to promote and increase awareness of environmentally and socially sound business practices. This includes determining the appropriateness of the sources of energy and water along with their usage efficiency, pollution control and waste management. Second, it aims at conserving the natural resource base on which tourism depends. Finally, it aims at improving the overall standards of the tourism industry (Table 2). The socio-cultural and socio-economic criteria, equity, fairness and community development are addressed. Thus, according to Fennell (2006, p. 215), 'what makes these accreditation programmes ethical is their focus on trying to bring about sustainable tourism through the regulations that they hold to be true'.

The objective of the eco-labels is to uphold responsible tourism practices through the promotion of environmental, social and economic values. Certification by EK is done on a purely voluntary basis where hotels, game lodges and tented camps take part. The eco-rating scheme has three levels of certification: Bronze, Silver and Gold in ascending order. Each one of these levels defines conditions for qualification. Gold eco-rating, which has been awarded to 10 establishments, is an indicator that a facility has achieved superior and replicable levels of excellence in responsible resource use, environmental conservation and socio-economic investment. Silver eco-rating, given to 58 establishments, demonstrates a move towards excellence; it is awarded to facilities that show innovation in responsible resource use, environmental conservation and socio-economic investment. The final category is that of Bronze eco-rating which currently has been awarded to 32 establishments. This category is the entry level and demonstrates a facility's awareness and commitment to environmental conservation, responsible resource use and socio-economic investment.

It is worth noting that among the Gold rated establishments is Turtle Bay Beach Club in Watamu (145 rooms with a bed capacity of 300) that describes itself as 'Kenya's number one all-inclusive resort' (www.turtlebay.co.ke). Of the total rated facilities, 44 (i.e. 44%) are located in Masai Mara and its environs. Future research should be done to determine the impact of such measures with regard to customers' decision making.

Table 1. Eco-rating certification fee.

Level	Application fees US$	Certification fee US$	Validity period
Gold	150	1350	2 years
Silver	150	850	2 years
Bronze	150	350	2 years

Source: www.ecotourismkenya.org

Figure 1. Eco-rating certification emblems.
Source: www.ecotourismkenya.org.

Towards a broader conceptualization of ecotourism

According to The International Ecotourism Society (TIES), ecotourism is defined as 'responsible travel to natural areas the conserves the environment, sustains the well being of the local people, and involves interpretation and education' (www.ecotourism.org). Moreover, one of the tenets of ecotourism is that it is supposed to be locally oriented and small scale (Fennell, 1999). However, with time there has been a tendency to broaden the scope of ecotourism. For example, Weaver (2003) advocated for the concept of 'mass ecotourism', noting that ecotourism can only be beneficial if it is brought to the masses. In the same way, EK advocates for the very concept of ecotourism to be adopted by a larger number of actors and facilities. It considers the conceptualization of ecotourism as a low volume restrictive activity. For this reason, it carried out the eco-certification of tourist enclaves, notably Turtle Bay Beach Resort having 145 rooms in the coastal town of Watamu (Gold rating), Baobab Beach Resort and Spa in Diani Beach (Silver rating), Serena Beach hotel and spa (164 rooms) in Shanzu area (Silver rating), and that of Lake Nakuru lodge within the Lake Nakuru National Park (Bronze rating), among others.

It is clear that EK views ecotourism as a facilitator that brings together communities, tourism and conservation. However, as with Fennell's (1999) *helicopter ecotourism* in Hollywood which does not involve close contact with nature, it is not easy to notice linkages between some of the eco-rated establishments and the local communities who live nearby. For such *helicopter ecotourism*, it is not easy to see the link between 'conservation, communities, and sustainable travel' as exemplified by TIES motto (www.ecotourism.org). Is this a case where a good idea such as eco-certification is adopted but deployed superficially and then subsequently ruined by altering its use? Indeed, Fennell (1999, p. 272) posited, 'Let the large mega resorts and mass tourism become sustainable tourism, *not ecotourism*'. This implies that such facilities as Turtle Bay practice *sustainable mass tourism* rather than ecotourism in Fennell's (1999) interpretation of the word.

It is therefore apparent that the concept of ecotourism is still undergoing a metamorphosis. Over time, it has been applied indiscriminately to anything that establishes a link between tourism and nature (Fennell, 1999), such as the case of safari-based tourism in Kenyan parks. For some it is a philosophy, a frame of mind, a process rather than a type of tourism. Indeed, EK affirms that:

> ecotourism *is not a type of tourism* like other types of tourism e.g. adventure, beach, culture-based, wildlife, sports, scenic, specialized etc; it is not a comparison or alternative to mass tourism; and *it is not an exclusive way of doing tourism.* (www.ecotourismkenya.org)

It notes that Ecotourism is 'a tourism business approach or strategy for achieving the goals of

Table 2. Considered parameters in the evaluation criteria in Kenya.

Environmental criteria	Waste management criteria	Resource management	Purchasing and supplies	Health and safety	Social cultural criteria
Environmental management	Grey water management	Energy management	Environmental consideration/ waste reduction	Environmental health and safety	Training and education
Environmental conservation	Black water management	Water management	Chemical use	Occupational health and safety	Benefits to staff and local community
	Solid waste management				Culture preservation
Visitor education	Waste oil				Visitor communication and education
	Pollution				Staff communication and education

Source: Personal communication from eco-rating department of EK.

sustainable development. Hence ecotourism principles are applicable to all types of tourism and tourism products' (www.ecotourismkenya.org).

Thus, what was once seen as 'alternative' and 'minority' has come to be the mainstream in today's tourism, for example, extension of the concept to urban areas. However, it is generally agreed that such tourism can only take place in 'natural areas' (www.ecotourism.org) – an antonym of urban areas. This is termed as 'urban ecotourism' (Wu, Wang, & Ho, 2010). Nonetheless, with EK the concept has been stretched to the beach with 13 beach establishments having been certified as meeting the eco-rating criteria. This means that *new* ideas such as ecotourism are now beginning to be used in traditional settings such as the beach (*beach ecotourism*?)

It is worth noting that this paper does not make value judgements as to whether such enterprises are meeting the standards for ecotourism or not. As Honey (2008, p. 113) rightfully puts it, 'ecotourism is going to move from a good concept to good practices'. If this is the intention of EK, then it presents an opportunity to stretch the concept of ecotourism further into non-traditional areas and settings. Nevertheless, at least it should be noted that establishments, especially those on the Kenyan coast, do not really portray the ecotourism image. Being all-inclusive and/or enclaves, there is minimal contact between their clientele and the local populations. Honey (2008) observed that mass tourism resorts have been using the 'eco' label, sometimes with little or no substance.

However, eco-labels project an image of a responsible facility (Sallows & Font, 2004). For EK, the aim is to 'recognize best practices in environmental conservation, responsible resource use and socio-economic investment among tourism accommodation facilities' (www.ecotourismkenya.org). The negative ecological footprint of tourism justified the award of labels by EK without discrimination to make the industry sustainable, responsible and competitive while moving ecotourism from exclusivity to inclusivity.

Ecotourism in the context of sustainability

One of the concerns has been the small number of tourists paying for the 'eco-labelled' products due to premium pricing (Fennell, 2006). Sallows and Font (2004, p. 98) noted that, 'eco-labels are meaningless in the majority of the cases and tourists perceive that certified products are more expensive … [and] are not widely recognized or understood by the average tourist'. Thus, while certification can be seen as a beneficial tool if properly implemented, by 2010 there were only 50 eco-rated establishments in Kenya. This number rose by 100% to reach 100 establishments under this scheme by the year 2012.

On a national scale, there are over 10.5 million hotel rooms available in Kenya as at 2011 (GoK, 2012a). With only 100 establishments eco-rated, the concern has been the low uptake of such a process by the accommodation sector in Kenya. Thus, for the largest part of the Kenyan tourism sector ecotourism has not worked. Even on the international scale, Sallows and Font (2004) noted that less than 1% of all tourism operations are involved in some form of certification. Nevertheless, such a small number of eco-rated facilities can hardly make an impact at a macro level since the eco-certification process has not been widely embraced. For example, in Nairobi (Kenya's capital city) area, there is only one facility with 23 rooms that has been eco-rated. Further research should be done to ascertain the real benefits of eco-rated facilities from the business perspective and its accrued advantages to the service providers. Have such facilities managed to create clear brand recognition in order to divert customer attention from the rival option of non-eco-rated facilities? With such an approach an eco-rating scheme will increase in value and be embraced by the larger accommodation sector.

However, it is noted that the aim of eco-certification is: to promote and increase awareness of environmentally and socially sound practices; to promote the increased contribution of tourism to socio-cultural development; to promote the increased contribution of tourism to conservation of natural resources upon which tourism is dependent and to improve destination image. With only 100 eco-rated establishments and 440 members since inception, the whole idea of eco-rating by EK has not been taken seriously by the industry.

Thus, while the process of certification in Kenya shows dynamism and a transformation in general, little seems to have changed. Kihima (2014b, p. 56) observed that, 'the contemporary issue has been to attract the masses, to lodge them and to take them round with little being done towards ensuring environmental integrity'. Thus, the environmental challenges such as off-road driving and animal harassment attributed to the Kenyan tourism in the 1970s are still being witnessed today. The good intentions are rarely translated into good practices. The motives of different establishments being involved in the certification process are not also clear (Fennell, 2006). In addition, as noted by Sallows and Font (2004), certification can increase inequalities by promoting only those who have the capacity to fulfil assessment requirements (Table 1). This is true in the Kenyan scenario since few, if any, truly or real community based and owned projects (for example, the Kakamega community bandas) are eco-rated by EK.

Linking tourism to communities: Kakamega forest bandas

In 2003, as one of its major milestones, EK initiated the Isecheno Banda Project. This project is situated in an under-exploited tourism destination of Kakamega Forest Reserve despite its description as 'the canopy of natural beauty' (Kihima, 2010). This forest, measuring 240 km^2, is the only remnant of a rainforest in Kenya offering great but unrealized tourism potential (Okello, Wishitemi, & Mwinzi, 2001), including but not limited to bird and primate watching, butterfly watching and education (research).

Despite such noble intentions towards fulfilling the mission of EK of 'linking tourism, communities and conservation for sustainable development in Kenya', it is observed that in most cases, destination areas are vast with only one or a few entrances to the natural area. Tourism therefore has had an impact on only a few people around the natural resource base. The implication is that there is no intimate connection between the reserve and the surrounding community, hence no or little benefit to the local people. Around this forest, there are 35,000 homes, each having an average of 4.8 people (GoK, 2007). This translates into approximately 140,000 people, yet tourism activity is restricted to the villages close to main entrance of Isecheno in the South and Buyangu in the North (Kihima, 2014a). This disfranchises most members of the community in destination areas in so far as tourism benefits are concerned. A survey undertaken by Kihima (2010) among local community members revealed that there is a considerable variation between those who suffer negative effects (82%) such as human wildlife conflict and those benefiting from the tourism activity (28%) in Kakamega area ($N = 100$). This therefore presents a challenge of linking communities, tourism and conservation.

Thus, in as much as ecotourism is ecologically based, it proves difficult to ensure a close relationship between the natural environment, the tourists and the local communities, especially if the conservation area is vast in size. In essence, tourism revenue from wildlife is limited to small portions of protected areas. Consequently, the challenge lies in increasing the benefits of tourism beyond the vicinity of the park entrance areas (Kihima, 2014a). Indeed, while the negative impacts attributed to the natural environments (the damage caused by wild animals, denial of accessibility, congestion) are experienced at a local level, the positive impacts (the economic benefits and job creation) are global or national. Therefore, the conflicts around biodiversity areas persist because such territories project an international image often disconnected from local development (Kihima, 2010).

Low volume, low impact in ecotourism projects

In addition, the emphasis on a small number of visitors in ecotourism definitions makes it even unviable to support local communities, as in the case of Kakamega forest bandas. This is more so in cases where clients to such areas are not high end. The Kakamega forest bandas constitute only six Isecheno bandas with a total bed capacity of 18. Compounded by fewer numbers of visitors (Table 3) and benefit sharing challenges at a community level, Manyara and Jones (2007) concluded that CBOs have little impact with regard to poverty reduction. The revenues in Table 3 are meagre given that the average modern wage employment for the country is US$572 (Ksh. 63 = 1 US$, 2007 exchange rate) per month (GoK, 2012b). Therefore, given that such CBOs tend to rely on ecotourism, it should be noted that solving one problem (environmental degradation) by encouraging small-scale projects and low visitor numbers may create another problem of low economic benefits for the local communities. Certainly, fewer tourists in destination areas are not synonymous with ecotourism, neither is ecotourism restricted to low visitor numbers.

Indeed, the main challenge is not the setting up of community projects but rather the key concern should be the viability of such projects in the long run as well as the integration of such projects in the mainstream tourism industry. It is evident that there is no tourism without the tourist. This therefore implies that such initiatives must be cognizant of the benefits gained from tourist information by communicating to potential

Table 3. The number of visitors recorded in the community bandas of Kakamega in 2007.

Month	Visitors	Revenue in US$ (December 2007 exchange rate of Ksh. 63 to the dollar).
January	64	984
February	62	694
March	21	190
April	62	786
May	24	175
June	37	333
July	100	1075
August	85	Not available
September	39	Not available
October	26	Not available
November	33	Not available
December	38	Not available
Total	591	7937 (approximation)

Source: Kihima (2010).

clients (Gunn, 1994). Boniface (2001, p. 39) recommended that, 'as providers and consumers, we need to know in order to respond to change'. EK whose vision is 'to be a leader in the knowledge and practice of ecotourism' should have done more by preparing the local communities for the tourism industry. This can be realized through the provision of a proper marketing platform for the CBOs (Community Based Organizations) since at the very basic level such initiatives must be known to exist. According to Boniface (2001, p. 43):

> information is pivotal to the tourism industry in order to deliver a versatile, viable product, as well as to enable a market to understand the use and care of the item that the industry has provided in response to the market's perceived wishes.

However, this is not normally the case in that sometimes community members owning these enterprises would wait for tourists to come without really being pro-actively involved. For example, the community bandas at Isecheno do not have any promotional messages and materials such as brochures targeting visitors. The email address is hardly used, no signposts are visible and no community member is in charge of marketing. Information and Communication Technology (ICT) has not been fully embraced with the website information link on www.ecotourismkenya.org under 'community groups/Kakamega Environmental Education Programme' lacking in appeal and presentation. Indeed, one of the key missing elements of these projects is lack of openness and visibility to the tourism markets.

Diffusing ecotourism activities at a local level

Even for the tourists at Kakamega, the forest is the principal attraction. Around this forest, tourism is principally nature-based with little interest shown towards the local surroundings and culture (Kihima, 2014a). Here, the tourists visit the forest without being interested in village tours. There is no concrete interaction between the tourists and the local communities (Kihima, 2010). This may be due to the fact that there are low levels of interest or capacity within the local community to serve the tourism industry within their locality, or tourism may be an industry that is too intricate for them to handle. The relationship must be such that the tourists should be fascinated by the cultural aspect (if provided by the local people) and that the local people should seize the opportunity to benefit from such visits (Kihima, 2014a). However, this is not always the case. For example, in Bigondi village in Uganda, Lepp (2008, p. 1210) noted that despite the tourists' interest in dining with the local people, 'not a single resident has made an effort to capitalize on the apparent market for roasted goat and dinners with locals. Instead, the residents are content to wait passively for their name to be called by a passing tourist'. According to Lepp (2008, p. 1210) this implied that 'many residents simply wait for chance of good fortune to land on them rather than actively pursue it'.

Indeed, the interaction between the locals and the tourists in Kakamega area is superficial, through gestures in the form of 'greetings' and within a 'framework of research' but not in the strict aspect of 'doing tourism' (Kihima, 2010). The gap between the local communities and the tourism sector puts to test the EK motto. Wheeller (1997) summed it up by noting that sustainability is an exercise in futility if local communities are not actively involved through participation and eventual attainment of benefits.

The involvement of members of the local community in conservation and tourism must go beyond simple initiation of such projects, which lack concrete structures for their implementation and management. These types of projects such as the bandas are far from being integrated in the host community, such that very few people benefit from them. For example, there are only 40 'active' members (EK records 200 members on their website) in this group, while the nearby community is much larger. Moreover, a good majority of community members as well as the local director of the Kenya Forest Service (KFS) denounce this association for having no impact on the rest of the community (Kihima, 2010). Thus, without careful follow-up over the years these projects become more disconnected from the community. They become exclusive even though in theory they are termed as projects done *for, by* and *with* the host community. This may be attributed to lack of motivation among local populations, the majority of whom are not competent in tourism related activities as well as the fear of remaining indefinitely volunteers without any accrued benefits. However, even if such ventures give hope to the local people to engage in conservation, it shows weaknesses inherent in local participation. Most frequently the community members are not self-organizing or skilled enough to build initiatives. Their progress depends heavily on outside intervention and financial backing.

Conclusions

The aforementioned presentation shows the determination by EK in ensuring that nature is conserved rather than exploited carelessly (Table 2). Although EK has tried to lead the way towards sustainability, the debate surrounding the concept of ecotourism is still fraught with paradoxes and gaps. First, it has revolved around the well-being of the natural environment by being described as a soft, responsible and non-perturbing displacement. For example, it is prohibited to exceed the speed limit of 40 km/h and to drive off the designated tracks within the Kenyan terrestrial parks, not to crowd the animals or make sudden noises or movements, and not to feed the animals among other regulations required of travellers.

Nevertheless, popular representation of ecotourism as 'responsible travel to natural areas … ' as defined by The International Ecotourism Society (TIES) seems to forget the role played by the public sector stakeholders in maintaining the integrity of the environment. A requirement for responsible travel is described by Prosser (1992, p. 39) as being rhetoric in the sense that:

> when on holiday … we may give little thought to whether such behaviour is socially or environmentally acceptable in our chosen destination, and indeed, if we are aware, we may rationalize that we have paid and they have taken our money, so we are entitled to enjoy our holidays as we wish.

It is therefore unlikely that tourists will be 'working' on ensuring that their holiday behaviour is directed towards the tenets of ecotourism. Being an escape activity, tourists would not want to be immersed in the worries of the day-to-day life such as environmental protection. This means that many people express a sense of entitlement to enjoy their holidays as they choose (Miller, Rathouse, Scarles, Holmes, & Tribe, 2010). Nevertheless, in many instances just like the KWS wildlife code, the focus is mainly on the traveller whom we expect to travel responsibly, ethically and 'softly'. Even EK defines ecotourism as:

> The *involvement of tourists* in environmental conservation activities directly linked to addressing human development needs, and promotes equitable sharing of benefits accrued from tourism with local communities while supporting their nature conservation values.

However, beyond the 'responsible traveller', a wider reflection is needed, which takes into account 'responsible development and planning' for concerned territories. Gunn (1994, p. 3) emphasized that tourism is not inherently destructive; rather, 'the truth is that tourism development is being done by those who have too little understanding of what tourism really is and how to plan and design it'. This shows that it is not only the tourist who has the responsibility of improving the environment through his or her activities (Fennell, 2006), but also deliberate actions through tourism planning especially at site level are needed (Gunn, 1994).

Certainly, the genesis of the negative tourism impacts does not lie in the very act of touring but in *tourismification*. While the attraction needs of the travellers must be integrated in places such as parks, the ultimate responsibility lies with destination managers through planning. This is so because tourism spaces provide service functions, including but not limited to accommodation, recreational facilities, accessibility, watch towers and communication equipment among others. Thus, beyond just responsible travellers (as seen in the code of conduct proposed by the KWS in protected areas), responsible hoteliers and tour guides (as certified by EK), other stakeholders such as KWS and local authorities in tourism also need to act responsibly, especially in destination planning and development. Without them the destination would not be sustainable. For example, although the majority of certified establishments by EK are in the Masai Mara region, the irony is that it is in this very ecosystem that there continues to be serious signs of degradation (Muriithi, 2014). Mere certification cannot save the scenery or disguise a bad policy for implementation of good tourism plans and development.

Second, the call for dynamism within the tourism industry is necessitated because 'tourism, as it stands, is already not working very satisfactorily for many involved in it, whether as a tourist or provider' (Boniface, 2001, p. 4). However, 'the essence of dynamism is not merely to change, but to be in a constant state of change' (Boniface, 2001, p. 14). While the move to undertake an eco-rating scheme as well as set up community based projects is laudable due to the reasons aforementioned, it is critical to note that such isolated measures without concerted efforts from all stakeholders may come to nothing. There seems to be a gap in monitoring outcomes of such initiatives. There is still a need to find the reason behind low participation in the eco-rating schemes by the accommodation sector in Kenya.

Finally, it should be noted that the competitive dimension of the tourist destination calls for vigilance among public authorities. In many instances, the basic problem resides in the policy of laissez-faire with regard to the management of destination areas, institutional failure, and selective nature of the local administration to respect the existing laws, politicization of the natural environment issues and also high levels of corruption. While eco-certification may not be able to change the current state of affairs, it is a step towards making the micro destinations in the country competitive, hence promoting the macro destination. Undoubtedly, even for Kakamega Forest Reserve ecotourism can thrive at a significant level. On the other hand, this destination must decide to make ecotourism a privileged choice by effectively linking community projects and the tourism market.

Disclosure statement

No potential conflict of interest was reported by the author.

References

Boniface, P. (2001). *Dynamic tourism, journeying with change*. Clevedon: Channel View Publications.

Cater, E. (1994). *Ecotourism: A sustainable option*. Chichester: John Wiley and Sons.

Fennell, D. A. (1999). *Ecotourism, an introduction*. London: Routledge.

Fennell, D. A. (2006). *Tourism ethics*. Clevedon: Channel View Publications.

Fyall, A., & Garrod, B. (2005). *Tourism marketing: A collaborative approach*. Clevedon: Channel View Publications.

Government of Kenya. (2007). *Economic survey 2006*. Nairobi: Kenya National Bureau of Statistics.

Government of Kenya. (2012a). *Kenya facts and figures 2012*. Nairobi: Kenya National Bureau of Statistics.

Government of Kenya. (2012b). *Economic survey*. Nairobi: Kenya National Bureau of Statistics.

Government of Kenya. (2014). *Economic survey 2014*. Nairobi: Kenya National Bureau of Statistics.

Gray, E. G. (2004). *Doing research in the real world*. London: Sage Publications Inc.

Gunn, C. A. (1994). *Tourism planning. Basics, concepts and cases* (3rd ed.). Washington: Taylor and sons.

Honey, M. (2008). *Ecotourism and sustainable development: Who owns paradise*. Washington, DC: Island press.

Hoteliers' blatant disregard of law puts mara ecosystem at risk. (2010, February 18). *Daily Nation Newspaper*.

Kenya Institute of Public Policy Research and Analysis. (2009). *Kenya economic report 2009, building a globally competitive economy*. Nairobi: KIPPRA.

Kihima, B. O. (2010). *Towards unlocking tourism potential in less visited destinations in Kenya: The case of Western Kenya Region* (Unpublished PhD. Thesis). University of Lyon 2, France.

Kihima, B. O. (2014a). People and parks in Kenya: toward Co-existence? The case of Saiwa Swamp and Kakamega forest natural areas. *Sociology Study, 4*(2), 170–181.

Kihima, B. O. (2014b). Unlocking the Kenyan tourism potential through the park branding exercise. *Tourism Recreation Research, 39*(1), 51–64.

Lee, J-S., Hsu, L-T., Han, H., & Kim, Y. (2010). Understanding how consumers view green hotels: How a hotel's green image influence behavioral intentions. *Journal of Sustainable Tourism, 18*(7), 901–914.

Lepp, A. (2008). Tourism and dependency: An analysis of Bigondi village, Uganda. *Tourism Management, 29*, 1206–1214.

Manyara, G., & Jones, E. (2007). Community-Based tourism enterprises development in Kenya: An exploration of their potential as avenues of poverty reduction. *Journal of Sustainable Tourism, 15*(6), 628–644.

Mayaka, M. A., & Prasad, H. (2012). Tourism in Kenya: An analysis of strategic issues and challenges. *Tourism Management Perspectives, 1*(2012), 48–56.

Meguro, T., & Inoue, M. (2011). Conservation goals betrayed by the uses of wildlife benefits in community based conservation: The case of Kimana Sabtuary in Southern Kenya. *Human Dimensions of Wildlife: An International Journal, 16* (1), 30–44.

Miller, G., Rathouse, K., Scarles, C., Holmes, K., & Tribe, J. (2010). Public understanding of sustainable tourism. *Annals of Tourism Research, 37*(3), 627–645.

Muriithi, J. K. (2014). The role of certification in enhancing achievement of sustainable ecotourism development in Kenya. In J. Rieucau (Ed.), *Le Kenya dans la 3e Révolution Touristique* (pp. 153–181). Paris: L'Harmattan.

Odunga, P. O. (2005). *Choice of attractions, expenditure and satisfaction of international tourists to Kenya* (Unpublished Ph.D. Thesis). Wagenigen University, Netherlands.

Okello, M. M., Wishitemi, B. E., & Mwinzi, A. M. (2001). Relative importance of conservation areas in Kenya based on diverse tourist attractions. *The Journal of Tourism Studies, 12*(1), 39–49.

Ondicho, T. G. (2012). Local communities and ecotourism development in Kimana, Kenya. *Journal of Tourism, XIII*(1), 41–60.

Ospina, S., & Parry, K. (2004). Qualitative research. In G. Goethals, G. Sorenson, & J. Burns (Eds.), *Encyclopedia of leadership* (pp. 1280–1285). Thousand Oaks, CA: SAGE Publicataions.

Pearce, P. L. (2005). *Tourist behaviour: Themes and conceptual schemes*. Clevedon: Channel view publications.

Prosser, R. (1992). The ethics of tourism. In D.E. Cooper, & J.A. Palmer (Eds.), *The environment in question: Ethics and global issues*. (pp. 37–50) London: Routledge.

Sallows, M., & Font, X. (2004). Ecotourism certification criteria and procedures: Implication for Ecotourism management. In D. Diamantis (Ed.), *Ecotourism: Management and assessment* (pp. 90–109). London: Thomson.

Sarantakos, S. (2013). *Social research* (4th ed.). New York: Palgrave Macmillan.

Weaver, D. B. (2003). *Ecotourism in the less developed world*. Wallingford: CABI publishing.

Wheeller, B. (1997). Here we go, here we go, here we go eco. In M.J. Stabler (Ed.), *Tourism and sustainability, principles to practice* (pp. 39–49). Wallingford: CAB International.

Wishitemi, B. E. L., & Okello, M. M. (2003). Application of the protected landscape model in Southern Kenya. *Parks, 13*(2), 12–21.

World Resource Institute; Department of Resource Surveys and Remote Sensing, Ministry of Environment and Natural Resources, Kenya; Central Bureau of Statistics, Ministry of Planning and National Development, Kenya; International Livestock Research Institute. (2007). *Nature's benefits in Kenya, An atlas of ecosystems and human well-being*, Washington, DC: World Resources Institute.

Wu, Y-Y., Wang, H-L., & Ho, Y. F. (2010). Urban ecotourism: Defining and assessing dimensions using fuzzy number construction. *Tourism Management, 31*, 739–743.

Academia and the evolution of ecotourism

Bob Mckercher

ABSTRACT

This review paper traces the evolution of ecotourism and the role academia has played in that evolution. Ecotourism arguably would not exist, and certainly would not exist in the form it is now, were it not for the active involvement of the academic community. Academics and academia have played a critical stewardship role in the conceptualization and operationalization of this activity, more so than any other form of tourism. Ecotourism has evolved, and is still evolving through three phases. The 'New Dawn' phase is typified by idealism, hyperbole and hope. It is followed quickly by a 'Crisis of Legitimacy', where critics illustrate that the ecotourism largely fails to meet its social, economic and ecological ideals, that has not yet been resolved fully. Finally, ecotourism reaches maturity when it achieves the 'Sustainable Product Niche' phase where a realistic understanding of what it can and cannot do, emerges.

Introduction

Ecotourism has been around in one form or another for about 30 years. Beginning as a western concept, the idea has now gained global appeal, with virtually every destination region offering some type of 'eco' experience. It is unique among tourism phenomena, though, for its conceptualization and operationalization is intrinsically tied to the evolving academic discourse on the subject. The link is so strong that Buckley (2008) noted that information sent to consumers about ecotourism by the commercial tourism press is largely consistent with and based on academics' beliefs about the activity. No other tourism sector can claim such a close relationship, if for no other reason than they were well established before the proliferation of tourism programmes in the 1980s.

Ecotourism and its academic discourse have evolved through three phases. Phase one can be titled 'the New Dawn' phase and is typified by idealism, hope, hyperbole and wishful thinking. Naïveté and mistakes made by both industry and academia here quickly led to 'a Crisis of Legitimacy' phase with the realization that many of the original ideals have not been met and that, indeed, may never be met. Ecotourism was subjected to much criticism, self reflection and, in many cases, condemnation. However, this stage was necessary to propel the sector to a third, more mature phase where ecotourism is recognized as a 'Sustainable Niche Product' typified by a realistic understanding of what it can and cannot do, and how effectively its idealistic claims can be measured. This review paper examines this evolutionary path through a discussion of the major themes of ideology/philosophy, market understanding, product development and delivery/business viability and, ultimately, sustainability. The author then explores the role academia has played in shaping the sector.

Ideology, Definition and Philosophy of Ecotourism

Parks Canada published an *Ecotour of the Rideau Canal,* in 1978. This 16-page guidebook may represent the first public use of the term. The booklet set out to enhance visitors' enjoyment of the Rideau Canal and to help users understand how human and natural history are interrelated (PC 1978). It was Hector Ceballos-Lascuráin, however, who is generally credited with popularizing the idea when, in 1983, he described a new form of nature travel that placed greater emphasis on experience, learning and emerging environmentalism (Morgan 1999). The idea gained popular support almost immediately (Ceballos-Lascuráin 1993) due to the convergence of growing environmentalism, a backlash against mass tourism and the alleged fracturing of the mass market.

It, along with the broader issue of sustainability, became dominant themes in the tourism literature throughout much of the last 25 years, due also to the convergence of a number of conditions, including the proliferation of tourism programmes, the creation of 'new' universities in many jurisdictions, a shift in focus of tourism education programmes and the resultant contested nature of tourism education, the quest for tourism to be recognized as a legitimate academic field of study, and the academic background of tourism academics. This role will be referred to throughout the paper, and then discussed in detail in the last section.

Sustainability entered the popular lexicon in the 1980s with the publication of the 'Bruntland Report', more accurately titled *Our Common Future* (WCED 1987). This report drew attention to the fact that the traditional approaches to economic development were not ecologically or socially sustainable in the long run. Fundamental change was required, therefore, in how we think of development to ensure a future that meets the needs of the present without compromising the ability of future generations to meet their own needs.

While sounding simple, in theory, it soon became a contested idea in practice, as two competing schools of thought vied for dominance. The development-oriented approach argued on behalf of the 'constant wealth concept', whereby wealth is calculated based on the aggregate value of both natural and man-made capital. The total asset base should grow between generations, although the absolute stock of natural assets may be permitted to decline providing that its use generates wealth (Pearce *et al.* 1989). Two fundamental principles drove this interpretation. The first is that the natural resource base can be consumed, degraded or otherwise used providing it produces wealth. The second principle suggests that optimum resource utilization should drive decision-making. Thus, activities that produce less wealth can and should be displaced by those that produce more wealth.

The ecologically sustainable perspective, or 'constant natural capital' approach on the other hand, adopted a more cautious approach, arguing that the natural asset base cannot be permitted to decline through time (Hare 1990; Pearce *et al.* 1989). Unlike the constant capital approach, built assets will be permitted to grow only with the proviso that they do not degrade the absolute stock of natural assets. This approach argues that species biodiversity, ecosystem integrity, non-substitutability of certain activities, uncertainty and the threat of irreversibility of impacts must take precedence in any development decision. The controversy still remains largely unresolved as conservationists, ecologists and climate specialists argue for constant natural capital, while industry lobbyists and governments in emerging and/or stagnating economies opt for the constant wealth approach.

Tourism was identified by both groups as one activity where sustainability could be achieved. It was portrayed as a smokestack-free industry that relied on selling of pristine or near pristine environments as its core product. As such, an assumption was made that the tourism industry had a natural interest in preserving the resource on which it relied. Additionally, tourism was also touted as a viable economic opportunity for peripheral regions that had few other options available (NES 1994). Much of the rhetoric at the time also revolved around how tourism could be a powerful force to reduce poverty, preserve traditional lifestyles and reinvigorate traditional cultures (NES 1994; Western 1993). It also coincided with the propagation of the belief that somehow indigenous people had a stronger conservation ethic than their Western counterparts (Fennell 2008) and, by extension, much could be learned by adapting mass tourism around indigenous land management principles.

Ecotourism quickly emerged as face of sustainable tourism and was portrayed as a new type of tourism for a new type of socially conscious consumer (Hawkins and Lamoureux 2001; Wight 2001). This belief was fuelled in part by research suggesting that the mass market was fracturing (Poon 2003) as tourists rejected traditional holidays and instead were looking for more meaningful experiences (Remington 1994). A substantial number of proponents even believed that this new group of tourists would soon displace the traditional tourist and emerge as the new mainstream market (Litvin 1996). Terms such as anti-tourist, alternative tourist, responsible tourist, green tourist and others were promoted to describe this emerging market (Cazes 1989). None gained broad acceptance, except 'ecotourism' which quickly became the universal expression of forms of alternative tourism.

In retrospect, many of the ideals were based more on wishful thinking than fact. Ideology soon clashed with reality, plunging ecotourism into a crisis of legitimacy that it has still not fully emerged from. Empirical evidence revealed that many of the espoused ideals were not met, nor could they ever be met (Higham 2007). Economic benefits were usually smaller and more highly localized than anticipated (Lindberg *et al.* 1996). Income in developing world eco-destinations proved to be sporadic, seasonal and concentrated among a few individuals. Inter-community differences were accentuated by gender, political and patronage inequities (Belsky 1999). Actions by NGOs to impose western business practices on indigenous communities were well intentioned, but often had limited effect (Barking 2003). The introduction of commercial ecotourism often hindered ecological preservation by creating a strong backlash against conservation (Belsky 1999; Barkin 2003). Unexpected financial costs were generated, in addition to social and economic costs (Lindberg *et al.* 1996), leading some people to question the net benefit of these programmes. Quality concerns affected the sector almost immediately as it became clear that many businesses were using the label without any commitment to any of ecotourism's principles (WTO 2002a).

Ecotourism itself became a deeply contested issue. No common definition could be agreed upon, even though general concepts of sustainability, nature-based and educational benefits were inherent in most attempts (Donohoe and Needham 2006). Weaver (2001) illustrated that more than 85 different definitions existed, with many being value driven. Perhaps such confusion is to be expected with any emerging activity, but definitional and conceptual challenges became especially problematic when the term was misappropriated by industry and government as a marketing tool with little commitment to any of its ideals. At its worst, ecotourism became little more than politically correct mass tourism. Wight (1993) noted how eco-sellers were exploiting consumers' interests in green products by packaging experiences as ecotourism that had little to do with ecotourism, while Orams (1995:5) commented 'in some cases ecotourism may well be nothing more than a new marketing gimmick which dresses up existing tourism attractions in an attempt to increase market share.'

The debate became polarized between eco-generalists and eco-purists. The generalists adopted an inclusive view and saw it as part of the broader nature-based spectrum, but with a stronger emphasis on environment, ecology and education (Diamantis 1999; Herbig and O'Hara 1997). The more exclusive purist view held that true ecotourism had to focus exclusive on the environment, ecology and education and that any business, by definition, had to operate at a small scale to minimize its impacts. Financial returns should be a secondary objective to conservation (Malloy and Fennell 1998). Yet, ironically, adherents of this group also were the strongest proponents of the belief that ecotourism could be targeted at a mass market. Unfortunately, there was little empirical evidence to support the idealistic claims of purists. Malloy and Fennell (1998) reviewing the literature at the time, noted that while a common view existed that 'true' ecotourism organizations should behave in a morally superior manner to other businesses, they found no evidence to support this claim.

The inability to achieve both ecological and economic goals, though, was the greatest challenge faced by the sector. This issue was the core thesis of an excellent paper written by Lawrence *et al.* (1997:307), who commented: 'its potential is currently limited by a problem of legitimacy growing out of the tension between economic and environmental goals; a problem which is exacerbated by the relative newness of the industry'. Ecotourism was subjected to a great deal of criticism in the media, with headlines in Australia such as 'Be a Tourist, Visit Beautiful Places and Destroy Them' (Schwartz and Nicholson-Lord 1990), 'Loving the Reef to Death' (Mellor 1990) and 'Loving our Parks to Death: a cautionary tale' (Wescott 1993) and similar condemnations from some members of the academic community.

The causes of the crisis of legitimacy can be attributed to a number of factors. Many academics truly believed the values inherent in ecotourism were so powerful that the inherent rightness of ecotourism philosophically and ideologically was so strong that it could overcome any obstacles to its successful implementation. Because it was so good, its imposition could be justified anywhere. Conservationists, though, argued that some places were simply too fragile for any form of tourism, eco or otherwise, creating conflict between them and tourism proponents. Legitimate fears were also raised that ecotourism could initiate the development life-cycle in remote areas, and might, therefore, prove to be the most destructive form of tourism (Lawrence *et al.* 1997). The assumption that local residents and indigenous people instinctively had a stronger environmental ethic than others, seems to be a long-standing belief that also is not supported by fact. Fennell's (2008) excellent paper questioning the myth of indigenous stewardship illustrates that it began with Christopher Columbus and grew along with the myth of the noble savage. Yet, he illustrates that evidence from biologists, archeologists and anthropologists does not support the myth. Additionally, it was extremely naïve to assume that ecotourism could replace mass tourism and that most people wanted to change their travel styles. This issue will be discussed further in the following section of the paper.

The response by industry and many in the academic community was not to reconsider the ideological and philosophical foundation on which some of these false assumptions were based, but instead to call for the creation of various certification programmes, ecolabels and formal educational programmes to enforce the ideals. The exact number of such schemes is unknown, but Buckley's (2002a) review identified various studies that set the figure at between 60 and more than 100. The motivation behind their creation was ostensibly to improve standards, ensure quality, and assist in marketing

(Buckley 2002b), but more cynical motives of revenue generation and establishment of accrediting bodies as peak agencies were also observed. Most programmes proved ineffectual (Zhong *et al.* 2007) as low awareness levels and limited impact on purchase decisions meant that the hoped-for marketing benefits did not accrue (Fairweather *et al* 2005; WTO 2002a). Moreover, little evidence exists that they have improved ecological performance, raising legitimate questions about their efficacy (Buckley 2002a, b). Accredited properties did not achieve higher standards than non-accredited ones (Warnken *et al.* 2005).

A number of universities also began to offer ecotourism degrees. Observation and personal experience indicate that some of these early programmes were well-thought out, but that others were initiated opportunistically with the goal of securing student numbers, rather than providing high quality education. Aspects of existing parks and recreation and business tourism programmes were merged to create new degrees with minimal specialist ecotourism content and few new expert staff hired. Enrolments initially were high, but few graduates seemed to be able to secure career-developing jobs due to the small and fragmented nature of the industry.

But crises are never enduring, and out of this crisis of confidence emerged some clarity about the real nature of ecotourism, with the generalist view prevailing. Whereas ecotourism was conceptualized initially as operating outside of and apart from mainstream tourism, both ideologically and spatially, today it is seen as a complementary activity to tourism, and indeed as an activity that is often reliant on a resort-based consumer as its core market (Kontogeoropolous 2006). Current ecotourism thought places it firmly as a part of tourism, and as an activity that fits within the broader nature-based sector (Blamey 2001). At the same time, the belief of ecotourism entering the tourism mainstream and displacing other forms of tourism has been abandoned (Anon. 2003).

Proscriptive definitions are declining in popularity with the realization that ecotourism can now be pursued in a variety of urban, suburban, peripheral and natural area settings and where a spectrum of experiences can be had (Weaver 2002a, 2005). Soft ecotourism focuses more on entertainment and recreation with an edutainment element. Mega fauna and sightseeing are the main product attributes. A deep or comprehensive view seeks to provide a more holistic understanding of the functioning of ecosystems and is grounded in providing a deeper learning experience as Weaver (2002a) illustrates.

Evolving Market Understanding

Understanding of the ecotourism market has also progressed through similar stages. Once thought to be large and almost unlimited in growth potential, it is accepted now that it is a small niche market (WTO 2002a, b). How small? Ecotourist package holidays represent less than one per cent of the German outbound market (WTO 2002c). Only about four per cent of US outbound travel involves some element of nature-based tourism, but not necessarily ecotourism (WTO 2002d). These figures are a far cry from first generation research conducted in the late 1980s and early 1990s repeated frequently by academics that indicated 60% of all international tourists are nature-based tourists and 40% are wildlife-related tourists (IES 2000). It represents a significant change from findings reported by Herbig and O'Hara (1997) who indicated eco and adventure tourism represents 10% of international travel and were growing at 30% per annum. Other studies set the market size as ranging from US$ 10 billion to US$ 100 billion and with the market share varying from two per cent to 13% of the global market, as reported by Weaver (1999). Importantly, it also challenges UNWTO's own studies made as recently as 2000, that reported that 20% of global travel involved ecotourism (as cited by Chin *et al.* 2000).

How could early researches have been so wrong, and how could the academic community have accepted such findings without question for so long? The answer, again, lies in a combination of the reliance on small scale, unsophisticated, and often methodologically flawed research, coupled with the fact that much of this early research was conducted by enthusiasts who let their beliefs bias objectivity. "New Dawn" research was often based on small, statistically unreliable samples. It sought to quantify the size of the market and its underlying motives but treated it as an undifferentiated mass. Some of the earliest motivational research confused motives with place features (Eagles 1992), while others failed to distinguish between ecotourism, adventure tourism and nature-based tourism, leading one researcher to suggest the US National Forests will generate US$ 3.5 billion from timber sales and US$ 100 billion from (eco)recreation (Mitchell 1997 cited in Morgan 1999). Other studies were based on very small and unrepresentative samples. Weaver (1999:795) concluded that much of the early estimates on the market 'emerged from the extrapolation and subsequent dissemination of anecdotal data concerning the performance of certain US specialty travel agents.'

Much of this research was based on the discredited secondary-data analysis method of analysing activity questions in departing visitor surveys to infer trip purpose and, therefore, market size. McKercher and Chan (2005) discuss this issue at length. They highlight how the size of the market can be over-estimated by up to a factor of 20 times when activity questions that measure what people did in a destination, are interpreted as motivational questions inferring trip purpose. Blamey (1995) and others illustrated as early as in the mid-1990s that a fundamental difference exists between someone who visits a protected area for any reason and a person who travels to a country specifically to pursue ecotourism. He showed what can happen when activity questions are misinterpreted by demonstrating that about 50% of tourists to Australia visit national parks, while only about five per cent came here specifically for that purpose (Blamey 1995).

The same method was and is used to identify differences between ecotourists and mainstream tourists which also led to the promulgation of many unfounded myths about ecotourists possessing a number of attractive features such as staying longer, spending more and being more active within the destination (Wight 1996; Eagles 1992). An early Canadian study attempted to document differences between ecotourists and mainstream tourists, based on a sample of 120 Canadians who traveled to Kenya (Ballantine and Eagles 1994). While differences were noted, they have more to do with the discriminating effect of distance on visitor profile than on any underlying 'eco' premise. But, again McKercher and Chan (2005) question the validity of this method at length. They raise three key issues. First, since activity questions cannot be used as proxies for trip purpose questions, then one cannot compare participants in one activity to non-participants to draw conclusions that draw inferences that the activity 'caused' people to stay longer in a destination. Second, they raise the practical issue that this method does not control the moderating effect of all other activities pursued during the trip that may also play a role in extended stays, etc. At its heart, though, this method attempts to draw causal conclusions by measuring one effect of visitation (what people did) with other effects of visitation (how much they spent, what else they did, how long they stayed, etc.). No causal relationship can be assumed. Indeed, visiting a site does not induce longer stays any more than wearing pajamas induces people to eat more breakfast cereal. A relationship may appear to exist, but the true cause can be attributed to more mundane factors, such as time of day for cereal consumption and pajama wearing or in the case of tourism, being a first-time visitor, staying at a main destination, participating in packaged tours, and the like. Indeed, most ecotourism activities in multi-product destinations represent secondary attractions that are pursued opportunistically, rather than as a core element of a trip. This realization explains the establishment of many ecotourism activities in or adjacent to large urban or resort destinations, as noted by Kontogeoropolous (2006).

Such unrealistic market estimates, quickly induced another crisis of legitimacy, only this time it was focused on commercial viability of eco businesses. The market history of ecotourism, like many other forms of special interest tourism, has been dominated by phantom demand; demand that various market research studies 'prove', but that does not translate into consumers who are willing to purchase products. A large number of studies conducted in various locales have discussed the marginal business viability of the sector, due primarily to unrealistic estimates of the potential market size (Pearce and Wilson 1995; Litvin 1996; Acott *et al.* 1998; McKercher and Robbins 1998; Loon and Polakow 2001; De Haas 2002). The UNWTO report at the World Ecotourism Summit noted that commercial viability of ecotourism initiatives was a recurring theme across most jurisdictions (WTO 2002a).

Fortunately, more sophisticated and less biased research began to appear that is providing more reliable figures. These studies have also identified different shades of green tourists, and in doing so, began to situate the ecotourism product by primacy of trip purpose and depth of experience desired. Pearce and Wilson (1995) were among the first to raise the prospect of an ecotourist continuum. Their study of wildlife-viewing tourists in New Zealand concluded that only a small minority identified wildlife-viewing as the primary motive for visiting the destination, while most indicated it was a lower order motive. As a result, a large majority of people purchased their tour on the day. The authors concluded that many visitors have only a casual interest in wildlife viewing. Tao *et al.* (2004) came to similar conclusions when examining the Asian ecotourism sector.

Weaver (2002a) presents an excellent review of segmentation studies that identified the core ecotourism market as being extremely small. His own study of guests at an ecolodge in an Australian national park found that fewer than 8% of the clients could be classified as 'hard core' ecotourists. Acott *et al.* (1998) evaluated the differences between deep and shallow ecotourism and deep and shallow ecotourists. By extension, the shallow ecotourist sees the environment as a resource which can be exploited to maximize personal benefits, without consideration of intrinsic value. They note further that shallow ecotourism verges on mass tourism where 'the actual tourist experience would be modified through a cultural window where western standards of safety, comfort and hygiene are maintained wherever possible. Although the shallow ecotourist might be interested in learning about the local culture, it would be from a distance, while maintaining familiar, immediate situations and circumstances. There would be no attempt at a deeper identification with landscapes or culture' (Acott *et al.* 1998: 245).

Product Viability

Naïveté about the market was accompanied by equally unrealistic assumptions about the skill sets required to run businesses and the practicalities of remaining commercially viable. Again, as with the market, early understanding of products was based on beliefs that almost any rural setting had ecotourism potential and that virtually anyone had or could develop the skills needed to run successful businesses (NES 1994). Indeed, a prevalent belief existed that somehow ecotourism products could defy traditional business wisdom, in much the same way that the 'dotcom' boom was seen to herald a new economy where traditional rules did not apply.

The academic community must accept some responsibility for fostering such myths, for many early generation ecotourism academics did not have a strong business background, even though they may have been housed in business schools. Initial optimism was driven by a number of perceptions: that demand was growing faster than the ability of supply to keep up with it; that mass tourists would pay a premium for quality green experiences; that low barriers to entry meant that capitalization was not an issue; that locally owned businesses would have an inherent competitive and ethical advantage over non-local firms; and that small-scale operations were the ideal business model (Lück 2002).

But this optimism was soon challenged by the harsh realities of running businesses (WTO 2002a). Indeed, the small amount of academic literature on business practice highlights a rather disheartening legacy of commercial non-viability, especially in remote regions and in the developing world. Most new ventures have a high failure rate, but the failure rate among first generation ecotourism businesses appeared to be even higher, often 90% or more (Anon. 2003). Loon and Palakow (2001) assert that, without government subsidization the risk associated with investing in southern Africa is too great to invest large amounts of capital into larger scale ecotourism projects. The De Haas' (2002) study of ecotourism in Nuie concluded that it may be culturally and ecologically sustainable, but not commercially viable due to insufficient visitor arrivals. McKercher and Robbins' (1998) study of Australian operators identified a number of business challenges facing the sector.

Low financial barriers may exist, but it also meant that far too many businesses entered the sector grossly undercapitalized and with no chance of ever being viable. These businesses faced severe cash flow problems from the outset and many were not able to withstand cash demands of the inevitable slower-than-expected growth. Marginal viability meant many businesses had to trade-off green practices for commercial survival. Operators did not appreciate the cost structure of micro businesses and the fact that they had to deal with high fixed costs per consumer. As such, prices had to be high, which in turn narrowed the potential market significantly. A lack of business acumen in general and marketing expertise in particular was common. Operators did not understand how to price their product, adopting a cost plus price, rather than pricing it on a value-plus basis. More so, though, their lack of knowledge of the tourism distribution system meant that few were willing to use the travel trade to reach the market, again limiting their market base (WTO 2002a).

The importance of location and setting became even more important as the sector matured (Wight 1996). A simple rule of thumb in tourism is that the more remote a product is from the source market, the smaller the potential market and the more unique it has to be in order to overcome the decaying of distance on demand. "Me too" products located in remote areas are unlikely to succeed if consumers can purchase a similar quality experience closer to home. History shows that isolated businesses will struggle, while locations where a critical mass of like businesses exist, tend to thrive. Here, success breeds success. Eco businesses now tend to be concentrated in certain geographic nodes that possess desirable market and landscape features (Dickey and Higham 2005; Weaver 2002b). The work by Deng *et al.* (2002) identifies key spatial and infrastructural attributes of the need for a wide variety of nearby cultural and natural resources, scientific value, rarity and aesthetic values, suitable facilities in and around the natural area, other attractions, accessibility, and community support as critical factors that must be in place.

But, again, out of crisis comes clarity. Those ventures that survived and learned, in spite of their mistakes, have played a critical mentoring role for aspiring new business people and offer valuable lessons for academic research on micro-business viability. A more strategic approach to business planning and development is also evident. Whereas once government departments used to publish 'how to manuals', specialist business planning texts began to emerge in the late 1990s (McKercher 1998). Weaver and Lawton (2007) have extended this research through the development of an excellent model illustrating the complex interrelationship between external conditions, supply and demand factors and the need to mitigate or minimize adverse impacts that is vital to success. Merging these conditions enables operators to provide suitable interpretative experiences for the appropriate market and product, effective marketing communication to transmit this desired mix to the consumer that sets expectations about the type of experience to be delivered. Successful businesses need to be well-financed and well-managed, integrated with other tourism businesses throughout the destination region and make strong use of global distribution networks and websites to reach the potential tourists.

Sustainability

Last, but not the least, the issue of sustainability has emerged as an on-going challenge, if for no other reason than the range of issues continues to evolve faster than ecotourism itself. Whether ecotourism could ever have achieved its lofty ideals is debatable, although optimists feel it can generally move in that direction (Lück 2002). Dimanche and Smith (1996) commented that tourism's environmental problems are not likely to be solved by ecotourism alone but by a general understanding of the tourism industry's responsibilities towards the environment. Yet, the dream of ecologically sustainable development was prevalent in the 'New Dawn' era. It was also promoted heavily by government departments that developed their own sustainable tourism development policies (ESDWG 1991). Ecotourism could make a substantial contribution to conservation, biodiversity and cultural heritage by providing an economic return and as a transformational activity that could change unsustainable land-use practices into more sustainable activities.

Ecotourism quickly fell victim to criticisms of empty rhetoric, with accusations of hypocritical attitudes about sustainability without enough definitive evidence, advice or guidance of how to convert the theory into practice (Morgan 1999). Perhaps such an accusation is unfair, as Higham and Carr (2002) illustrate that eco-experiences may provide an effective medium to influence environmental values. Certainly, examples exist where ecotourism has had a positive impact on natural resources by providing an alternative land use to more damaging activities and, indeed, in improving the overall quality of the natural asset (Li *et al.* 2006).

But, as Coburn-Isaacs (2000) argues, the potential of ecotourism as a wildlife conservation strategy is offset by its tendency to contribute directly to environmental degradation. Indeed, ecotourism was classified as, potentially, an insidious form of tourism for it introduced people into fragile areas often with little or no adequate planning (Shen *et al.* 2008) and, worse still, could trigger an influx of mass tourists as the area becomes more popular (Lück 2002). Recent research on carbon emissions also suggests that the carbon footprint of the long-haul ecotourist is substantially larger than that of the traditional short haul, resort based tourist (Becken 2002; Becken and Hay 2007). Simpson *et al.* (2008), for example, indicate that the carbon footprint of an ecotourist visiting the Antarctic is up to 1000 times greater than a bicycle tourist travelling domestically.

The crisis of legitimacy is especially prescient in emerging economies that have identified ecotourism as a regional development strategy (Stone and Wall 2004). Such places tend to adopt a constant wealth philosophy to sustainability and are more than willing to trade off natural values for economic development. Yet, community benefits may be limited and funds for conservation that are reliant on unstable tourism incomes may compromise the ability of destinations to maintain natural area integrity (Luo and Zheng 2008; Weaver 2005; Zhong *et al.* 2007). This issue is complicated by a lack of clarity over land ownership, open style management that focuses on income generation over conservation and the requirement that financing of nature reserves must come from income generated by ecotourism, which risks sacrificing conservation goals for income generation (Lindberg *et al.* 2003; Van Dijk 2006; Vanell 2008).

Part of the problem is that it is almost impossible to verify most of the claims about ecotourism's positive or negative effects on the environment, making it difficult to move from theory to practice. Buckley (2009) identifies four main barriers to effective scientific examination of its benefits and costs: lack of agreement on what is to be included in any aggregate evaluation of impacts; a lack of generally agreed parameters or accounting protocols to quantify and compare the various types of environmental costs and benefits; the complexity of many of the potential mechanisms to be used that must include ecological, social and political processes; and, finally, a lack of a critical mass of studies conducted using consistent methods across a range of products and activities to general preliminary impression of their environmental significance, on which to base future research.

While such criticism may be valid at some level, mature businesses and destinations also realize that any tourism exerts a variety of positive and negative impacts. Many small- and large-scale resorts are adopting best-practice technologies to minimize their impacts (WTO 2002a). The mature concept of sustainable ecotourism now embodies the 'triple bottom line' concept of seeking to optimize social, economic and ecological benefits. Commercial viability remains critical, but it needs to be associated with a true sense of corporate social responsibility. Still, though, sustainability remains an unrealized goal and arguably an unrealizable goal, for tourism cannot operate in isolation of other sectors. It is difficult for ecotourism to function sustainably when the world is enveloped by a thicket of unsustainability.

Academia's Role

At the outset of this paper, the author commented that ecotourism is unique in that the academic community has played a key role in its conceptualization, framing, organization and, indeed, operationalization. The dramatic expansion of university-level tourism programmes and emergence of ecotourism in the public consciousness are closely linked. Academics have led the ideological debate and established the framework to delineate it from other forms of nature-based tourism. Academics also played a central role in the formation of many ecotourism associations in the belief that industry and academia could develop the sector cooperatively. Academics, through consultancy and involvement as experts in many foreign aid projects, played a central role in the establishment of ecotourism in many jurisdictions in both the developed and developing world. And, finally, a number of academics were among the first to venture into business, especially in Australia. This last section presents a personal view of the complicit role of academics and academia in the successes and failures of ecotourism.

Both the number of universities offering tourism programmes and the focus of the tourism curriculum underwent seismic shifts beginning in the late 1980s. Expansion of programmes in the UK, Australia, New Zealand, Hong Kong and other countries influenced by the British system, came as a result of the elimination of the 'polytechnic' and 'colleges of advanced education' (CAE) educational tier through their conversion to full-fledged universities. The number of universities increased geometrically in a number of jurisdictions. Many, if not most, of these 'new' universities embraced tourism studies as part of their curriculum, while established universities rarely did so. The number of providers in Australia, for example, grew from three in 1987 to 16 by 1990 (Craig-Smith *et al.* 1995), to more than 25 today, with only one 'sandstone' university offering tourism as a result of a forced merger with a former agricultural college. Airey and Johnson (1999) noted similar expansion in the UK system, with undergraduate programmes expanding from two in 1986 to 66 by 1997. Ryan (1995), Goeldner (2001), Hobson (1999) and Hsu's (2005) edited text documented similar growth levels elsewhere.

The academic focus of these programmes also changed, from a traditional emphasis on vocational training to business studies, or as noted in many programmes in the United States from a peripheral study area in social science departments to a central field of study in business schools. In fact, almost all new tourism programmes adopted a business management model (Airey and Johnson 1999; Scott *et al.* 2008) used by other business disciplines, that incorporate a suite of general business management subjects with a core of specialist tourism subjects.

These changes fragmented the traditional tourism curriculum (Airey 2008), as various stakeholders with different agendas fought to have their specialist areas included. The degree of fragmentation, however, created a fertile breeding ground for the expansion of tourism research into a wide range of areas not examined previously, leading to the rapid expansion of the field both in terms of the quantum of research produced and also the breadth of topics covered. Ecotourism was one area to benefit the most.

Importantly, the supply of tourism academics capable of teaching business tourism did not keep pace with the growth of new programmes. Universities had to seek tourism specialists with other disciplinary backgrounds to staff their programmes. Many of these academics hired to teach in business-related programmes came from geography or recreation backgrounds and apparently conducted their research in a national parks or protected area management setting. About 30% of tourism PhDs granted between 1971 and 1987 came from geography or recreation (Jafari and Aaser 1988), while a follow-up study concluded that an even higher 39% of tourism dissertations completed between 1987 and 2000 were in these fields (Meyer-Arendt and Justice 2002). By comparison, the share of dissertations in economics fell from 35% to seven per cent during the same time, while the proportion of PhDs in the broad field of business administration stayed stable at less than 10%. A study of UK dissertations completed between 1990 and 1999

(Botteril *et al.* 2002) came to a similar conclusion, with 44% of dissertations completed in the broad topic area of tourism development or policy and planning, and only 13% examining tourism management at a firm or industry level. The dominance of geographers and recreationists in tourism scholarship was affirmed by Zhao and Ritchie (2007). Their study of 57 prolific scholars identified 31, or 54%, with doctorates in geography, leisure/recreation or forestry, seven in business management and three in economics.

This group of scholars needed to establish a role for themselves in non-business tourism areas or risk marginalization. Ecotourism was an obvious choice for many, for it provided an opportunity to apply their considerable skills within a relevant tourism framework. In theory and practice, ecotourism relates closely to protected area management, land-use planning, public land-use policy and outdoor recreation.

It is no surprise, then, that geographers and recreationists were among the first to engage in the discourse on ecotourism. They were largely responsible for establishing the field, and by extension their own ideological leanings and research interests also were complicit in plunging it into its crisis of confidence. The original purist view emerged from their collective belief that protected areas could serve both biocentric and anthropocentric needs, but with a stronger biocentric ethic. Further, many were ambivalent to commercial tourism, again as a reflection of prevailing attitudes towards acceptable uses of national parks and protected areas. Such biases are understandable given that most protected areas are reserved because of their special, unique or fragile characteristics. Moreover, their own lack of business backgrounds meant that this area of needed research was not high on their agenda. Instead, they had to accept the veracity of other studies unquestioningly, even though much of it was of varying quality. One also suspects that many repeated dubious, exaggerated figures as a reflection of their underlying belief in the concept of ecotourism and the hope that it could herald a new form of tourism.

The above comments are not meant to denigrate the early academic discourse of ecotourism, for it established it as a vibrant field of study. Criticism is easy in hindsight, but it must be remembered that these individuals were pioneers in an area where no previous knowledge existed. They were building knowledge as they progressed, with each piece of new research representing a significant advancement in the field. Basic knowledge is required before more sophisticated research can be undertaken, and undertaken it was. Figure 1 summarizes the number of full papers and review articles published on ecotourism (using ecotourism in the title, keyword or abstract) and the number of different journals that published these papers between 1991 and 2008. Data were derived from the Scopus abstract and citation data base (www.scopus.com) that covers nearly 18,000 peer reviewed journals. Only five papers were published each year in 1991 and 1992 and these were carried in three journals, two of which were *Annals of Tourism Research* and *Tourism Management*. Slow growth was noted to the mid-1990s when between 12 and 23 papers were published by between 7 and 13 journals each year. The suite of journals was limited largely to tourism and environmental science fields. Academic interest grew dramatically after 1995, with some 57 papers published in 41 journals in 1996. That figure has since grown to 200 papers in over 100 journals by 2008. The range of journals covering this topic has diversified, with geography, ecology, leisure, natural resource management and business journals beginning to publish papers in the field in the late 1990s, followed by rural studies, physical geography, animal behaviour, marine studies and a wide range of other journals published in all continents today.

Ironically, interest by the broader academic community may have initiated the ideological and sustainability crises of confidence faced by the sector and certainly

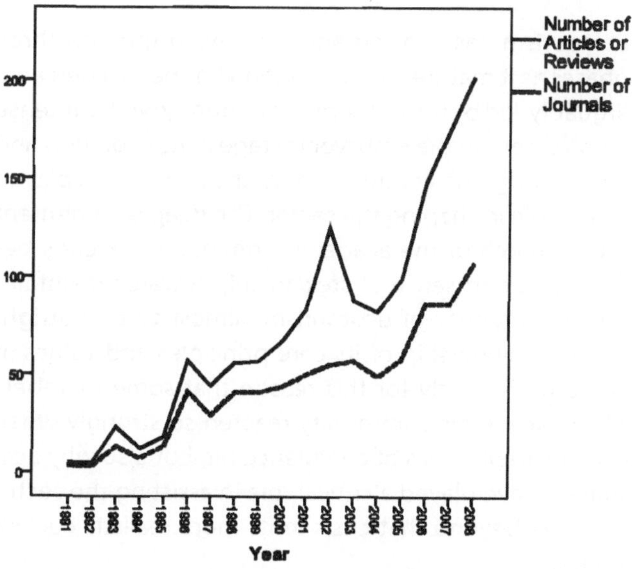

Figure 1. Number of Ecotourism Papers Published and Number of Journals Publishing these Papers 1991–2008.

Source: Scopus (Search for title, keyword and abstract).

was responsible for the crisis of product and market confidence noted previously. They have also played an instrumental role in the emergence from crisis to the 'Sustainable Product Niche' phase. Academics from non-tourism and recreation backgrounds redefined ecotourism and/or modified its core elements to suit their own research needs, embracing the simple initial concept in controversy. Further, as outsiders in tourism were not as committed to its ideals they could, therefore, initiate studies that challenged some prevailing myths. It is also evident that some scholars opposed the concept and undertook studies to refute some claims.

But these individuals also brought different skill sets that broadened the research base and enhanced the depth of examination. The resultant flourishing debate challenged some beliefs, reinforced others and ultimately led to a better understanding of what ecotourism is and is not. Importantly, a critical mass of knowledge was created, enabling researchers with an abiding interest in the topic to reflect more deeply on ecotourism, how best to deliver viable businesses, what its impacts were and how they could be mitigated and what the activity could and could not contribute to social, cultural and environmental sustainability. This depth of knowledge is responsible for the emergence of the 'Sustainable Product Niche' age we are now in.

Conclusion

Ecotourism has evolved and is evolving through three phases as it matures. This evolution has been aided and, arguably, led by the academic community whose intense involvement in the embryonic stage of its evolution and whose on-going commitment to the activity has played a vital role in shaping the sector. The deep commitment felt by much of the academic community extends beyond a strong sense of stewardship towards maintaining the integrity of ecotourism, almost to an outright sense of ownership of its core principles and values. It is, perhaps, partly for this reason that some members of the academic community reacted so strongly when the inevitable crisis of confidence hit, but also why academics have played a critical role in assisting the sector to move beyond that crisis to its more realistic current phase.

Ecotourism will remain an ever-evolving and, in some aspects, elusive activity. The term has suffered so widely from misuse that many in the sector avoid using the term and instead are reverting to such terms as sustainable tourism, eco-friendly tourism and low-impact tourism. But a market will always exists for this type of experience, regardless of its name. Some destinations will always seek to promote natural and cultural heritage as tourist attractions. Businesses will see an opportunity to match supply with demand. And sustainability will always be an issue.

References

ACOTT, T. G., LA TROBE, H. L. and HOWARD, S. H. (1998). An Evaluation of Deep Ecotourism and Shallow Ecotourism. *Journal of Sustainable Tourism* 6(3): 238–253.

AIREY, D. and JOHNSON, S. (1999). The Content of Tourism Degree Courses in the UK. *Tourism Management* 20(2): 229–235.

AIREY, D. (2008). Tourism Education: Life Begins at 40. http://epubs.surrey.ac.uk/cgi/viewcontent.cgi?article=1039&context=tourism – Accessed on 16 September 2009.

ANON. (2003). The EcoClub Interview – Ms Megan Epler Wood. *ECOCLUB* 5 (54) http://www.ecoclub.com/news/054/interview.html – Accessed on 7 May 2009.

BALLANTINE, J. L. and EAGLES, P. (1994). Defining Canadian Ecotourists. *Journal of Sustainable Tourism* 2(4): 210–214.

BARKIN, D. (2003). Alleviating Poverty Through Ecotourism: Promises and Reality in the Monarch Butterfly Reserve of Mexico. *Environment, Development and Sustainability* 5: 371–382.

BECKEN, S. (2002). Analysing International Tourist Flows to Estimate Energy Use Associated with Air Travel. *Journal of Sustainable Tourism* 10(2): 114–130.

BECKEN, S. and HAY, J. (2007). *Tourism and Climate Change: Risks and Opportunities.* Clevedon. Channel View Publications.

BELSKY, J. M. (1999). Misrepresenting Communities: The Politics of Community-based Rural Ecotourism in Gales Point Manatee, Belize. *Rural Sociology* 64: 641–666.

BLAMEY, R. (1995). Profiling the Ecotourism Market. In Richins, H., Richardson, J. and Crabtree, A. (Eds) *Taking the Next Steps.* Brisbane: Ecotourism Association of Australia: 1–9.

BLAMEY, R. (2001). Principles of Ecotourism. In Weaver, D. (Ed) *The Encyclopedia of Ecotourism.* New York. CABI: 5–22.

BOTTERIL, D., HAVE, C. and GALE, T. (2002). A Survey of Doctoral Theses Accepted by Universities in the UK and Ireland for Studies Related to Tourism, 1990–1999. *Tourist Studies* 2(3): 283–311.

BUCKLEY, R. (2002a). Tourism Ecocertification in the International Year of Ecotourism. *Journal of Ecotourism* 1(2/3): 197-203.

BUCKLEY, R. (2002b). Tourism Ecolabels. *Annals of Tourism Research* 29(1): 183–208.

BUCKLEY, R (2008). Testing Take-up of Academic Concepts in an Influential Commercial Tourism Publication. *Tourism Management* 29(4): 721–729.

BUCKLEY, R. (2009). Evaluating the Net Effects of Ecotourism on the Environment: A Framework, First Assessment and Future Research. *Journal of Sustainable Tourism* 17(6): 643–672.

CAZES, G. (1989). Alternative Tourism: Reflections on an Ambiguous Concept. In Singh T. V., Thuens, L. and Go, F. (Eds) *Towards Appropriate Tourism: The Case of Developing Countries.* New York. Peter Lang: 117–126.

CEBALLOS-LASCURÁIN, H. (1993). Ecotourism As a Worldwide Phenomenon. In Lindberg, K. and Hawkins, D. (Eds) *Ecotourism: A Guide for Planners and Managers.* Bennington. The Ecotourism Society: 12–14.

CHIN, C., MOORE, S., WALLINGTON, T. and DOWLING, R. (2000). Ecotourism in Bako National Park Borneo: Visitors' Perspectives on Environmental Impacts and their Management. *Journal of Sustainable Tourism* 8(1): 20–35.

COBURN-ISAACS, J. (2000). The Limited Potential of Ecotourism to Contribute to Wildlife Conservation. *Wildlife Society Bulletin* 28(1): 61–69.

CRAIG-SMITH, S., DAVIDSON, M. and FRENCH, C. (1995). Hospitality and Tourism Education in Australia: Challenges and Opportunities. In Faulkner, B., Fagence, M., Davidson, M. and Craig-Smith, S. (Eds) *Tourism Education National Conference Papers.* Canberra. Bureau of Tourism Research: 144–150.

DE HAAS, C. (2002). Sustainability of Small-Scale Ecotourism: The Case of Niue, South Pacific. *Current Issues in Tourism* 5(3&4): 319–337.

DENG, J., KING, B. and BAUER, T. (2002). Evaluating Natural Attractions for Tourism. *Annals of Tourism Research* 29(2): 422–434.

DIAMANTIS, D. (1999). The Concept of Ecotourism: Evolution and Trends. *Current Issues in Tourism* 2(2&3): 93–122.

DICKEY, A. and HIGHAM, J. (2005). A Spatial Analysis of Commercial Ecotourism Businesses in New Zealand: A 1999 Benchmarking Exercise Using GIS. *Tourism Geographies* 7(4): 373–388.

DIMANCHE, F. and SMITH, G. (1996). Is Ecotourism an Appropriate Answer to Tourism's Environmental Concerns? *Journal of Hospitality Marketing and Management* 3(4): 67–76.

DONOHOE, H. M. and NEEDHAM, R. D. (2006). Ecotourism: The Evolving Contemporary Definition. *Journal of Ecotourism* 5: 192–210.

EAGLES, P. (1992). The Travel Motivations of Canadian Ecotourists. *Journal of Travel Research* 31(2): 3–7.

ECOLOGICALLY SUSTAINABLE DEVELOPMENT WORKING GROUP (ESDWG) (1991). *Ecologically Sustainable Development- Draft Report Executive Summaries.* Canberra. Department of Arts, Sports, Environment, Tourism and Territories.

FAIRWEATHER, J., MASLIN, C. and SIMMONS, D. (2005). Environmental Values and Response to Ecolabels among International Visitors to New Zealand. *Journal of Sustainable Tourism* 13(1): 82–98.

FENNELL, D.A. (2008). Ecotourism and the Myth of Indigenous Stewardship. *Journal of Sustainable Tourism* 16(2): 129–149.

GOELDNER, C. R. (2001). Tourism Education: North American Experiences. *Journal of Teaching in Travel and Tourism* 1(1): 97–107.

HARE, M. (Ed) (1990). *Ecologically Sustainable Development: A Submission.* Australia. Australian Conservation Foundation, Greenpeace, The Wilderness Society and World Wide Fund for Nature-Australia.

HAWKINS, D. and LAMOUREUX, K. (2001). Global Growth and Magnitude of Ecotourism. In Weaver, D (Ed) *The Encyclopedia of Ecotourism.* New York. CABI: 63–72.

HERBIG, P and O'HARA, B. (1997). Ecotourism: A Guide for Marketers. *European Business Review* 97(5): 231-236.

HIGHAM, J. (Ed) (2007). *Critical Issues in Ecotourism: Understanding a Complex Tourism Phenomenon.* Burlington, USA. Butterworth-Heinemann.

HIGHAM J. and CARR, A. (2002). Ecotourism Visitor Experiences in Aotearoa / New Zealand: Challenging the Environmental Values of Visitors in Pursuit of Pro-Environmental Behavior. *Journal of Sustainable Tourism* 10(4): 277–294.

HOBSON, P. (1999). International Perspectives: A Comparison of U.S., U.K., and Australian Hospitality Education and Their Systems. In Burrows, C. and Bosselman, B. (Eds) *Hospitality Management Education.* Binghamton. Haworth Press: 213–238.

HSU, CATHY, H.C. (Ed) (2005). *Global Tourism Higher Education: Past, Present, and Future.* New York. The Haworth Hospitality Press.

INTERNATIONAL ECOTOURISM SOCIETY (IES) (2000). *Ecotourism Statistical Fact Sheet.* Washington. IES: 1–5.

JAFARI, J. and AASER, D. (1988). Tourism as the Subject of Doctoral Dissertations. *Annals of Tourism Research* 15: 407–429.

KONTOGEOROPOLOUS, N. (2006). Towards a Southeast Asian Model of Resort Based "Mass Ecotourism": Evidence from Phuket, Thailand and Bali, Indonesia, ASEAN *Journal of Hospitality and Tourism* 2: 1–16.

LAWRENCE, T., WICKINS, D. and PHILLIPS, N. (1997). Managing Legitimacy in Ecotourism. *Tourism Management* 18(5): 307–316.

LI, W., ZHANG, Q., LIU, C. and XUE, Q. (2006). Tourism's Impacts on Natural Resources: A Positive Case from China. *Environmental Management* 38: 572–579.

LINDBERG, K., ENRIQUEZ, J. and SPROULE, K. (1996). Ecotourism Questioned: Case Studies from Belize. *Annals of Tourism Research* 23: 543–562.

LINDBERG, K., TISDELL, C. and XUE, D. (2003). Ecotourism in China's Nature Reserves . In Lew, A., Yu, L., Ap, J. and Zhang, G. (Eds) *Tourism in China.* Binghamton. Haworth Press: 105–122.

LITVIN, S. (1996). Ecotourism: A Study of Purchase Proclivity. *Journal of Vacation Marketing* 3(1): 43–54.

LOON, R. M. and POLAKOW, D. (2001). Ecotourism Ventures: Rags or Riches?" *Annals of Tourism Research* 28(4): 892–907.

LÜCK, M. (2002). Looking into the Future of Ecotourism and Sustainable Tourism *Current Issues in Tourism* 5(3&4): 371–374.

LUO, J. C. and ZHENG, J. M. (2008). Ecotourism in Nature Reserves in China: Current Situation Problems and Solutions. *Forestry Studies in China* 10(2): 130–133.

MALLOY, D. and FENNELL, D. (1998). Ecotourism and Ethics: Moral Development and Organizational Cultures. *Journal of Travel Research* 36: 47–56.

MCKERCHER, B. (1998). *The Business of Nature-based Tourism.* Melbourne. Hospitality Press.

MCKERCHER, B. and ROBBINS, W. (1998). Business Development Issues Affecting Nature-Based Tourism Operators in Australia. *Journal of Sustainable Tourism* 6(2): 173–188.

MCKERCHER, B. and CHAN, A. (2005). How Special is Special Interest Tourism. *Journal of Travel Research* 44(1): 21–31.

MELLOR, B. (1990). Loving the Reef to Death? *Time Magazine* (November 5): 48–55.

MEYER-ARENDT, K. and JUSTICE, C. (2002). Tourism as the Subject of North American Doctoral Dissertations, 1987–2000. *Annals of Tourism Research* 29(4): 1171–1174.

MORGAN, D. (1999). Ecotourism – Fad or Future for Tourism? Number 18 – July 1999 http://www.bgci.org/education/article/371/ – Accessed on 15 September 2009.

NATIONAL ECOTOURISM STRATEGY (NES) (1994). *National Ecotourism Strategy Report.* Canberra. Australia Commonwealth Department of Tourism.

ORAMS, M. (1995). Towards a More Desirable Form of Ecotourism. *Tourism Management* 16(1): 3–8.

PARKS CANADA (PC) (1978). *Ecotour of the Rideau Canal.* Smiths Falls. Parks Canada: 16.

PEARCE, D., MARKANDYA, A. and BARBIER, E. B. (1989). *Blueprint for a Green Economy.* London. Earthscan Publications.

PEARCE, D. and WILSON, P. (1995). Wild-life Viewing Tourists in New Zealand. *Journal of Travel Research* 34(4): 19–26.

POON, A. (2003). Competitive Strategies for a New Tourism. In Cooper, C. (Ed) *Classic Reviews in Tourism.* Clevedon. Channel View Publications: 130-142.

REMINGTON, D. (1994). ecoTOURISM. *Profiles* 4(1): 16–19.

RYAN, C. (1995). Tourism Courses: A New Concern for New Times? *Tourism Management* 16(2): 97–100.

SCHWARTZ, L. and NICHOLSON-LORD, D. (1990). Be a Tourist, Visit Beautiful Places and Destroy Them. Melbourne. *The Sunday Age* (August 19).

SCOTT, N., PUELO, V. and CROTTS, J. (2008). An Analysis of Curriculum Requirements among Hospitality and Tourism Management Programmes in AACSB Colleges of Business in the United State. *Journal of Teaching in Travel and Tourism* 7(4): 71–83.

SHEN, X., SU, C. and LI, W. (2008). Jiuzhaigou Ecotourism Areal System: Temporal Evolution of Entropic Change. *Wuhan University Journal of Natural Sciences* 13(7): 303–308.

SIMPSON, M. C., GÖSSLING, S., SCOTT, D., HALL, C. M. and GLADIN, E. (2008). *Climate Change Adaptation and Mitigation in the Tourism Sector: Frameworks, Tools and Practices.* Paris, France. UNEP, University of Oxford, UNWTO, WMO.

STONE, M. and WALL, G (2004). Ecotourism and Community Development: Case Studies from Hainan, China. *Environmental Management* 33(1): 12–24.

TAO, C. H., EAGLES, P. and SMITH, S. (2004). Profiling Taiwanese Ecotourists Using a Self-definition Approach. *Journal of Sustainable Tourism* 12(4): 149–168.

VAN DIJK, J. D. (2006). Tourism and Ecotourism Development in China. Paper presented for P.R.E.M.I.U.M. Programme, Sponsored by the National Science Foundation and Michigan State University.

VANELL, T. (2008). Environmentalism in China. Brandeis University. 5 pp http://tessavenell.com/greening-beijing/docs/venell-environmentalism- in-china.pdf – Accessed on 16 September 2009.

WARNKEN, J., BRADLEY, M. and GUILDING, C. (2005). Eco-resorts vs Mainstream Accommodation Providers: An Investigation of the Viability of Benchmarking Environmental Performance. *Tourism Management* 26(3): 367–379.

WORLD COMMISSION ON ENVIRONMENT AND DEVELOPMENT (WCED) (1987). Our *Common Future.* Melbourne. Oxford University Press.

WEAVER, D. (1999). Magnitude of Ecotourism in Costa Rica and Kenya. *Annals of Tourism Research* 26(4): 792–816.

WEAVER, D. (Ed) (2001). *Encyclopedia of Ecotourism.* New York. CABI.

WEAVER, D. (2002a). Hard-core Ecotourists in Lamington National Park, Australia. *Journal of Ecotourism* 1(1): 19–35.

WEAVER, D. (2002b). Asian Ecotourism: Patterns and Themes. *Tourism Geographies* 4(2): 153–172.

WEAVER, D. (2005). Comprehensive and Minimalist Dimensions of Ecotourism. *Annals of Tourism Research* 32(2): 439–455.

WEAVER, D. and LAWTON, L. (2007). Twenty Years On: The State of Contemporary Ecotourism Research. *Tourism Management* 28(5): 1168–1179.

WESCOTT, G. (1993). Loving our Parks to Death: A Cautionary Tale. *Habitat Australia* (February): 14-19.

WESTERN, D. (1993). Defining Ecotourism. In Lindberg, K and Hawkins, D. (Eds) *Ecotourism: A Guide for Planners and Managers.* Bennington. The Ecotourism Society: 7–11.

WIGHT, P. (1993). Ecotourism: Ethics of Ecosell? *Journal of Travel Research* 31(3): 3–9.

WIGHT, P. (1996). North American Ecotourism Markets: Motivations, Preferences and Destinations. *Journal of Travel Research* 35(1): 3–10.

WIGHT, P. (2001). Ecotourists: Not a Homogenous Market Segment. In Weaver, D. (Ed) *The Encyclopedia of Ecotourism.* New York. CABI: 37–61.

WTO (2002a). *World Ecotourism Summit – Final Report.* Madrid. World Tourism Organization.

WTO (2002b). *The British Ecotourism Market.* Madrid. World Tourism Organization.

WTO (2002c). *The German Ecotourism Market.* Madrid. World Tourism Organization.

WTO (2002d). *The US Ecotourism Market.* Madrid. World Tourism Organization.

ZHAO, W. and RITCHIE, B. (2007). An Investigation of Academic Leadership in Tourism Research: 1985–2004. *Tourism Management* 28(2): 476–490.

ZHONG, L. S., BUCKLEY, R. and XIE, T. (2007). Chinese Perspectives on Tourism Eco-certification. Annals *of Tourism Research* 34(3): 808–811.

Submitted: May 12, 2009
Accepted: September 16, 2009.

Evolution of ecotourism: no crisis a response to McKercher (2010)

Ralf Buckley

McKercher (2010) argues that geographers (p. 22), who knew little about business (pp. 20, 23), held unrealistic hopes for ecotourism (pp. 16–19) which persuaded naïve entrepreneurs (pp. 15, 20) into undercapitalized start-ups (p. 20), and second-tier universities (p. 22) to enrol students in poorly-designed courses (p. 18) with no career prospects (p. 18); but then suffered a collective crisis of confidence (pp. 15–24), and have now retracted their projections (pp. 21–24). His arguments, however, suffer from a series of scholastic shortcomings, and his conclusions are not supported by the limited evidence actually available.

Some sections of his article are indeed accurate: for example, that tourism was touted as a 'smokestack-free industry' which could provide 'a wide economic opportunity for peripheral regions' (p.16); that income from tourism is often 'sporadic, seasonal and concentrated among a few individuals' (p. 16); that 'many businesses were using the [ecotourism] label without any commitment' (p. 17); that 'the term was misappropriated by industry and government as a marketing tool' (p. 17); and that ' [l]egitimate fears were also raised that ecotourism could initiate the development life-cycle in remote areas' (p.17). These components, however, do not support his central thesis.

Scholastic Shortcomings

McKercher's article is short on data. It does not provide economic statistics on historical trends or geographic patterns in ecotourism. It does not analyse the content of research publications, university courses, government policy documents, industry lobbying materials, commercial products, or marketing materials. It does not examine communications between academia and industry, either in person or through publications.

It makes assertions (pp. 16–19) about 'assumptions', 'rhetoric', 'beliefs', 'polarization', and similar phrases indicating knowledge of individual human mental states, with no evidence of any such knowledge. It claims (p. 22) that ecotourism research was established by academics who 'needed to establish a role for themselves in non-business tourism areas'; brought their own 'ideological leanings' to their research; were 'complicit' and 'biased'; 'had to accept the veracity of other studies unquestioningly'; and 'repeated dubious, exaggerated figures'. It claims repeatedly, but without evidence, that ecotourism has "plunged" into a 'crisis of legitimacy' (pp. 15–21), a 'crisis of confidence' (p. 23) or even an 'inevitable crisis of confidence' (p. 24).

It claims (p. 16) that ecotourism research arose from sustainable tourism. The earliest academic publications on ecotourism, however (Romeril 1985; Young 1986), were five years before the first on sustainable tourism (Nash and Butler 1990; Pigram 1990). Ecotourism research also has long antecedents in outdoor tourism and recreation, reviewed by Pearce (1985) and Edington and Edington (1986); whereas research in sustainable tourism did not gain momentum until the 1990s (Buckley 1996).

McKercher's article argues (p. 22) that academics 'played a central role in the formation of many ecotourism associations'. There are at least 16 ecotourism associations and societies worldwide[1]. For most, no academics are listed amongst the founders and Boards. The Australian association includes one academic in an 11-member Board. One of the founders of the Fijian association is an academic now, but not at the time.

In arguing for early exaggeration (p. 18), McKercher cites purported projections of US Forest Service ecotourism revenue. He takes these figures, however, from an unrefereed posting, with no citation data for the original reference, made by a government tourism official, on a botanical gardens website. In fact, these statistics are from the US Forest Service itself, are actual rather than projected, and refer principally to ski resorts.

In arguing for a high failure rate of ecotourism start-ups, McKercher (p. 20) cites a single consultant's opinion, from an interview in a non-academic newsletter, based on an uncited report, about international aid funding. McKercher and Robbins (1998), also mentioned, did not actually present data on failure rates for ecotourism business start-ups. Longevity data for Australian outdoor tourism operators (Buckley 1999) did not include survival or failure rates, since they were derived from a single retrospective survey. There are several known time-series datasets of ecotourism enterprises, but none have been analysed or published.

McKercher's article confuses cause and correlation, whilst criticizing other research for the same fault (p. 18–19). It confuses means and extremes in time series analysis. It confuses ecotourism, adventure tourism and nature-based tourism (pp. 18–19) whilst criticizing the same error.

It misquotes Weaver and Lawton (2007), Zhao and Ritchie (2007) and Buckley (2009a). It cites only selective and secondary sources in economics (p. 16), biology, anthropology and archaeology (p. 17). It draws sweeping generalizations from single case studies (pp. 16–21). It uses undefined and essentially meaningless terms such as "true" ecotourism and "true" corporate social responsibility; and tautologous terms such as "sustainable ecotourism". It concludes with a truism: "sustainability will always be an issue". Taken singly, these are not fatal errors; but taken together, they greatly weaken McKercher's contentions.

Evidence

McKercher's central thesis seems to be that both academic projections and actual commercial activity in ecotourism were small in the 1980s, larger in the 1990s, and smaller in the 2000s; and that industry trends were driven by academic involvement. To test these claims would require (i) a complete set or representative series of academic statements about the economic scale of ecotourism over the past three decades; (ii) a reliable three-decade time series on the number and scale of new commercial ecotourism businesses established, according to a standardized definition; and (iii) evidence of a causal link between academic research and commercial practice, i.e. that commercial tourism enterprises knew what academics were saying and changed their practices accordingly.

Recent reviews of publications and practices in ecotourism (Buckley 2009a,b; Weaver and Lawton 2007), make no suggestion of three historical phases, and neither do compendia of case studies (Buckley 2003a; Zeppel 2006). McKercher's article does not provides evidence to indicate otherwise.

Estimates of economic scale for ecotourism differ both because data are unreliable, and since they refer to different subsectors of the tourism industry. Tourists' actions don't match their motivations (McKercher 2010). Visitor counts on site don't match surveys at airports (Buckley 2003b, 2009c). Supply-side estimates don't match demand-side, and top-down estimates don't match bottom-up (Buckley 2009b, 2010a). Tourism enterprises which contribute to conserving biodiversity constitute only a tiny fraction of the outdoor tourism industry as a whole (Buckley 2009b, 2010a,b). None of this has changed. The number of reported nature-based tourism operations contributing to conservation and communities, however, has continued to increase slowly but steadily (Buckley 2010b; Gössling and Hultman 2006; NACSO 2008; Saarinen *et al.* 2009; Spenceley 2008; Stronza and Durham 2008).

There is little evidence of information flow from academia to industry. An ecotourism book by *Lonely Planet* drew only on secondary sources (Buckley 2008). The now-defunct Cooperative Research Centre for Sustainable Tourism, supposedly an academic-industry partnership, produced only one document with large-scale take-up: 20,000 copies of minimal-impact guidelines for four-wheel-drive tours (Buckley 2002). Academics and industry members of the Tourism and Travel Research Association 'have little awareness of each other in terms of knowledge production and use', according to Xiao and Smith (2010:410), who cited six previous publications with similar conclusions. Management agencies for World Heritage Areas consult both tourism practitioners and ecotourism researchers, but there is no evidence that the tour operators follow ecotourism research findings.

Conclusions

McKercher's article was not presented as hypotheses, autoethnography or personal opinion, but as a review. As outlined above, however, it does not provide primary data, and it mixes reputable refereed publications with the idiosyncratic opinions of individual industry protagonists, with little concern for methodological rigour. If the social sciences are ever to escape their current position at the bottom of the hierarchy of sciences (Fanelli 2010), they must, at the very least, demonstrate reliable academic scholarship.

McKercher claims (p. 15) that 'ecotourism arguably would not exist, and certainly would not exist in the form it is now, were it not for the active involvement of the academic community'. This seems to be simply incorrect. Historical trends in the economic scale of ecotourism, and the potential influence of academics, remain open questions. The weight of evidence, however, indicates firstly that commercial ecotourism is continuing to grow, slowly but steadily; and secondly, that academic research in ecotourism has had rather little influence on commercial practice.

Note

1. Australia, Asia-Pacific, Belize, Ecuador, Hawaii, Hong Kong, India, International, Kamchatka, Kenya, Malaysia, Murghab, Pakistan, Saskatchewan, Sweden, Toledo.

References

BUCKLEY, R.C. (1996). Sustainable Tourism: Technical Issues and Information Needs. *Annals of Tourism Research* 23(4): 925–928.

BUCKLEY, R.C. (1999). International and Australian Patterns in Nature, Eco and Adventure Tourism. *National Ecotourism Conference 1998*. Fraser Island. Ecotourism Australia.

BUCKLEY, R.C. (2002). Minimal-Impact Guidelines for Mountain Ecotours. *Tourism Recreation Research* 27(3): 35–40.

BUCKLEY, R.C. (2003a). *Case Studies in Ecotourism*. Wallingford. CAB International.

BUCKLEY, R.C. (2003b). Pay to Play in Parks: an Australian Policy Perspective on Visitor Fees in Public Protected Areas. *Journal of Sustainable Tourism* 11(1): 56–73.

BUCKLEY, R.C. (2008). Testing Take-Up of Academic Concepts in an Influential Commercial Tourism Publication. *Tourism Management* 29(4): 721–729.

BUCKLEY, R.C. (2009a). Evaluating the Net Effects of Ecotourism on the Environment: A Framework, First Assessment and Future Research. *Journal of Sustainable Tourism* 17(6): 643–672.

BUCKLEY, R.C. (2009b). *Ecotourism: Principles and Practices*. Wallingford. CAB International.

BUCKLEY, R.C. (2009c). Tourism in Parks. *PLoS Biol* 7(6): e1000143.

BUCKLEY, R.C. (2010a). *Adventure Tourism Management*. Oxford. Elsevier.

BUCKLEY, R.C. (2010b). *Conservation Tourism*. Wallingford. CAB International.

EDINGTON, J.M. and EDINGTON, M.A. (1986). *Ecology, Recreation and Tourism*. Cambridge. University Press.

FANELLI, D. (2010). "Positive" Results Increase down the Hierarchy of the Sciences. *PloS ONE* 5: e10068: 1–10.

GOSSLING, S. and HULTMAN, J. (2006). *Ecotourism in Scandinavia: Lessons in Theory and Practice*. Wallingford. CAB International.

MCKERCHER, B. (2010). Academia and the Evolution of Ecotourism. *Tourism Recreation Research* 35(1): 15-26.

MCKERCHER, B. and ROBBINS, B. (1998). Business Development Issues Affecting Nature-based Tourism Operators in Australia. *Journal of Sustainable Tourism* 6(2): 173–188.

NAMIBIAN ASSOCIATION OF CBNRM SUPPORT ORGANISATIONS (NACSO) (2008). *Namibia's Communal Conservancies: A Review of Progress and Challenges in 2007*. Windhoek. NACSO.

NASH, D. and BUTLER, R. (1990). Towards Sustainable Tourism. *Tourism Management* 11(3): 263–264.

PEARCE, D.G. (1985). Tourism and Environmental Research: A Review. *International Journal of Environmental Studies* 25: 247–255.

PIGRAM, J. (1990). Sustainable Tourism – Policy Considerations. *Journal of Tourism Studies* 1(2): 2–9.

PIGRAM, J.J. and JENKINS, J.M. (2006). *Outdoor Recreation Management*. London. Routledge.

ROMERIL, M. (1985). Tourism and the Environment: Towards a Symbiotic Relationship. *International Journal of Environmental Studies* 25: 215–218.

SAARINEN, J., BECKER, F., MANWA, H. and WILSON, D. (eds) (2009). *Sustainable Tourism in Southern Africa: Local Communities and Natural Resources in Transition*. Bristol. Channel View.

SPENCELEY, A. (2008). *Responsible Tourism: Critical Issues for Conservation and Development*. London. Earthscan.

STRONZA, A. and DURHAM, W.H. (eds) (2008). *Ecotourism and Conservation in the Americas*. Wallingford. CAB International.

XIAO, H-G. and SMITH, S.L.J. (2010). Professional Communication in an Applied Tourism Research Community. *Tourism Management* 31(3): 402–411.

WEAVER, D.B. and LAWTON, L.J. (2007). Twenty Years On: The State of Contemporary Ecotourism Research. *Tourism Management* 28(5): 1168–1179.

YOUNG, A.M. (1986). Eco-Enterprises: Eco-Tourism and Farming of Exotics in the Tropics. *Ambio* 15(6): 361–363.

ZEPPEL, H. (2006). *Indigenous Ecotourism: Sustainable Development and Management*. Wallingford. CAB International.

ZHAO, W. and RITCHIE, B. (2007). An Investigation of Academic Leadership in Tourism Research: 1985–2004. *Tourism Management* 28(2): 476–490.

Part III

Ecotourism practices and perspectives

Ecotourism, animals and ecocentrism: a re-examination of the Billfish debate

David A. Fennell

ABSTRACT

The aim of this paper is to discuss the topic of who, or what, is deserving of moral consideration in nature as this applies to the emerging discourse on tourism, animals and ethics. The paper focuses primarily on deep green or ecocentric ethics, and more specifically, on how this perspective differs from other theories of environmental ethics—especially animal liberation— when it comes to the treatment of animals used for tourism. The debate on billfishing as ecotourism is resurrected for the purpose of explaining (i) why it is important to understand key differences between ecocentric and animal liberation ethics; and (ii) how both theories could have been used to better inform the billfishing debate that took place over a decade ago. The paper concludes with a call for the continued use of environmental ethics theory in clarifying whether certain human-animal practices should be viewed as ecotourism or not.

Introduction

Environmental ethics emerged as a new sub-discipline of philosophy over 40 years ago, advancing on the back of work by the early pioneer Aldo Leopold (1970; first published in 1949), and galvanized later by Rachael Carson (1962), Lynn White (1967), and Garrett Hardin (1968). A chief objective of the discipline continues to be recognition of the importance of infusing moral theory into discussions on non-human entities (Desjardins 2013; see also Mehmetoglu, 2009). This has proven problematic because of the difficulty in deciding not only the reasons for preserving nature, but also what in the environment is deserving of moral standing (Ouderkirk 2000; Nash 1989; Humphrey 2000).

Questions surrounding moral value are paramount to the discipline because those entities possessing moral standing are owed certain duties, not unlike the duties that humans have towards each other. By contrast, those entities that fall outside the circle of morality are treated instrumentally, i.e., they have value only in as much as they serve the needs of humans. An early environmental visionary in this area was Rolston (1975, 1981), who argued that nature has intrinsic value alongwith a number of dimensions (e.g., economic, life support, recreational, scientific, aesthetic, life, and so on).

The lack of consensus on what falls inside the circle of morality and what falls outside has spawned many different theoretical perspectives that are organized according to scale as well as their focus on living and non-living entities. Examples of these include biocentrism, ecocentrism, the land ethic, deep ecology, eco-feminism, and animal liberation (see Desjardins 2013). These perspectives cover a range of important topics and related questions, including: Do individual animals have moral standing? Does the environment in general have moral standing? How is the moral standing that humans enjoy the same or different than the moral standing of animals and nature? Is it morally acceptable to kill animals?

The purpose of this paper is to discuss the source of moral value (moral standing) in reference to the emerging discussion on tourism, animals and ethics. In particular, the focus is on ecocentric ethics and, more specifically, on how ecocentrism differs from other kinds of environmental ethics, especially animal liberationist theories like animal rights and utilitarianism, when it comes to the treatment of animals used for tourism. The debate on billfishing as ecotourism is resurrected in the latter stages of the paper for the purpose of explaining: (1) why it is important to understand key differences between different environmental ethics theories like ecocentrism and animal liberation; and (2) how these theories could have been used to better inform the billfishing debate. This paper continues the recent trend of examining specific animal ethics theories and their application in a tourism context (Fennell 2012a; Yudina and Fennell 2013).

Environmental Ethics

Maturity in environmental ethics has come about through the development and refinement of a diverse range of theories that have been organized according to the degree in which they place emphasis on human- and ecologically-centred values. For the purposes of this paper, Curry's (2011) work on light green, mid-green and deep green categories of environmental ethics has been used to illustrate differences in these theoretical perspectives.

Light Green or Shallow Anthropocentric Ethics

The light green or shallow environmental ethic places direct value primarily on humans (Curry 2011). In support of this perspective, Norton (1984) argues that there are two basic forms of anthropocentrism, weak and strong, with weak anthropocentrism being sufficient enough to support an environmental ethic. An example of the weak perspective includes enlightened self-interest (see Payne 2010 in reference to anthropocentrism as human chauvinism), which reflects society's need to maintain a healthy environment simply for the purpose of maximizing human health and comfort (e.g., ecosystem services such as soil generation, water purification and oxygen production, according to Hull et al. 2003) today and for the future. We protect these services and treat them as resources simply because their healthy functioning is in our own best interests.

The ecological humanists, a group that includes social ecologists like Murray Bookchin (1990), fall within this shallow category of environmental ethics because they are unwilling to move beyond a constricted anthropocentric perspective (Humphrey 2000; see also Sessions 1995, and DiZerega 1992, 1995). This point is validated in consideration of the following two statements:

a. the human being is the sole object of value in the natural world;
b. the human being is the object of greatest value in the natural world.

Humphrey (2000) argues that ecological humanists would adopt at least one of these two value positions. Although there are certainly good reasons to protect various elements of nature, it is always the interests of humans that take precedence over the interests of these other entities (living or non-living). The humanist attempt to elevate humans as unique over all other beings is a way of legitimizing our domination over the planet and all of its inhabitants (Biro 2002; Eckersley 1992).

Mid-green or Intermediate Ethics

Mid-green environmental ethics is characterised as non-anthropocentric but not entirely ecocentric (Curry 2011). Value is not just placed on humans, but the extension of value also does not extend so far as to include ecosystems. The focus of the mid-green perspective is, as said by Curry (2011), to be more on animals, and more specifically individual animals. This type of ethics:

> *extends moral considerability to (primarily) animals, which are therefore perceived as possessing independent moral status, and therefore as deserving protection for their own sakes, regardless of whether they matter to human beings (Curry 2011: 72).*

Biocentrism and animal liberation are two main perspectives within this category. Biocentrism adopts the position that (1) all life has inherent value, and (2) this inherent value is held equally amongst all the constituent parts of nature. All living things, therefore, have their own 'good' (Taylor 1986), and one species cannot be viewed as being morally superior to another (French 1995). The locus of value is therefore placed not on the ecosystems in which these living entities exist, but rather on the entity itself. Moral standing and even rights are extended to individual organisms because these organisms are conscious, feeling and/or purposeful (Hull et al. 2003; see also Carter 2005). Hull and his colleagues continue by suggesting that this would include 'the right to life, freedom, and pain-free existence' (Hull et al. 2003:3).

Biocentrism can be criticized on the basis of inconsistency (French 1995). On one hand proponents argue for equality across the species, and on the other hand suggest that vital human interests can at times override the interests of non-humans. They 'toss species ranking out the front door of their arguments only to have it sneak around and into the house from the back' (p. 40).

Another main theoretical perspective of the mid-green environmental ethic is animal liberation, supported chiefly by the work of Peter Singer (2009) and Tom Regan (2004). These theories are said to be reductionist or atomistic in the way they defend the interests of individual animals—not species or whole systems. There is thus an unwillingness to sacrifice individual animals for the good of the whole (Regan 2004).

Utilitarians like Singer argue that moral standing or moral consideration is based on the individual, and this position is defended on the basis of equal consideration of interests, speciesism, and pain and suffering. This means that the many marginal interests that humans satisfy in using animals for pleasure or experimentation, should not outweigh the interests that these animals have in avoiding suffering. Any being, human or non-human, made to suffer more should be accorded a higher level of treatment over those suffering less (Fennell 2012c, for more on this theory).

Tom Regan's theory of animal rights extends moral consideration on the basis of the rights that animals have, not unlike the rights that humans enjoy. Animals that are seen to possess inherent value (normal mammals of one year or more in age and even fish, and birds) are said to be subjects-of- a-life based on the capacity for consciousness, intentionally, and sentience, and should never be treated as a means to an end. In linking his theory to justice, Regan contends that we have a direct duty to protect the interests of others, and this includes moral agents (humans) as well as moral patients (animals) (see Fennell 2012a, for an extended discussion of this theory).

Environmentalists have criticized animal liberation because there is no special place for endangered animal species as deserving special treatment because of their status. Instead, the focus rests solely on the good of individual animals. For example, in making a decision on how best to advance with one of two construction projects— one that would harm an endangered species and the other that would harm a non-threatened or endangered species—there is little assistance (Ouderkirk 2000).

Deep Green or Deep Ecocentric Ethics

While the mid-green perspective places more emphasis on the value of individual animals, ecocentric ethics places value on the whole systems. Examples of this perspective include the self-development and self-realization components of deep ecology (Naess 1984; Fox 1989), the Gaia hypothesis (Lovelock 1979), the land ethic (Leopold 1970), and by extension, the ethic of care and responsiveness of ecofeminism (Donovan 1990; Gruen 2010; Fennell and Sheppard 2011; Adams 1993; Yudina and Fennell 2013).

Ouderkirk (2000) illustrates that 'ecocentric' refers to the fact that 'in environmental matters the ecosystem must be a central consideration' (p. 24). Ecocentrists assign ecosystems with intrinsic value because they are self-organizing systems that generate benefits for flora and fauna, and we can never separate humans from these systems. Ecocentrism differs from biocentrism in the way the former ascribes intrinsic value to the whole, including those living and non-living parts, while the latter, as above, ascribes intrinsic value only to those living elements of the environment Eckersley (1992) writes that the world from the ecocentric perspective:

> *Is an intrinsically dynamic, interconnected web of relations in which there are no absolutely discrete entities and no absolute dividing lines between the living and the non-living, the animate and the inanimate, or the human and the non-human (p. 49).*

The consequentialist theorist, Aldo Leopold has had a significant impact on the way we think about ecocentrism. His sketches and observations on the land ethic took shape in the early 1930s (Leopold 1933a, 1933b) and culminated in his landmark work, A *Sand County Almanac,* in 1949 (Leopold 2004). His main maxim, i.e., 'A thing is right when it tends to preserve the integrity, stability, and beauty of the biotic community. It is wrong when it tends otherwise' (Leopold 1970:262), has been a mainstay for environmentalists since its publication. Leopold extends the boundaries of the moral community to include animals and plants, rocks, water, soil, and everything else that collectively comprises the land (Nelson 2004). As long as the consequences of our actions do not compromise the integrity of the biotic community, i.e., if we act in a way that promotes the good, and not harm, of the biotic community on the whole, these actions are morally praiseworthy (see Callicott, 1989 who proposed Leopold's land ethic as a global code, and Evanoff's (2007) critique of this view).

From a practical standpoint, as long as the ecosystem is respected, there is no limit to the types of activities that might take place in nature. If hunting, trapping, fishing and other consumptive activities are good for the whole system these activities would be deemed morally acceptable. (See Hettinger and Throop 1999, who feel that ecocentrisis ought to move away from integrity and stability, to more contemporary features like diversity and complexity.) In this capacity, Rowe (1997) argues that we ought to feel more pain at the devastation of wild ecosystems than over the death of individual organisms. This perspective has been adopted by Loftin (1985) who writes that it makes little sense to doctor sick animals because value should be placed not on individual animals but rather on the entire ecosystem. He argues that those who choose to shelter and aid these animals are biologically illiterate. Time, money and other resources would be better channelled into habitat protection or other measures that would protect the integrity of ecosystems.

Criticisms of Ecocentrism

Ecocentrism has been criticized along two related fronts. The first is the charge of an economic motive, i.e., that anthropocentric values guide ecocentric decision-making, and the second is based on environmental fascism. Both are discussed briefly, below.

For Rowe (1997), ecocentrism often incorporates too much cultural anthropocentrism and individual self-interest. He argues that if we start with individual humans as the locus of value in building concern for the ecosphere and work outwards to family, kin, community, and so on, i.e., working from the inside-out, instead of the outside-in, such will be the kiss of death for humanity and the world in general. Leopold, Rowe contends, is such a proponent of this method, espousing ecological well-being though Darwinian rationalizations, with air, water and other natural elements taking second place to the needs of humans and their immediate family and friends. Rowe argues that:

> *Aldo Leopold has been the influential exponent of ethics-by-extension, rationalized as an expedient for human survival. Unfortunately this approach only strengthens anthropocentrism, making it certain that land, air, water and other organisms will always in the crunch take second place to the welfare of self, family and friends (Rowe 1994: 2).*

Instead, Rowe (1997) favours the use of Feibleman's (1954) model of systems hierarchies, in particular, his 'laws of the levels', whereby the 'mechanism of any level is found at lower levels (in its parts), while the purpose of any level is found at levels above (in the wholes)' (p. 149). For Rowe, this means: 'Earth before organisms. Ecosystems before people. Ecosphere not biosphere. Ecocentrism not biocentrism'.

Lehman (1981) shares this view on Leopold in arguing that:

> *Although Leopold claims for communities of plants and animals a 'right to continued existence', his argument is homocentric, appealing to the human stake in preservation. Basically it is an argument from enlightened self-interest, where the self in question is not an individual human being but humanity -present and future — as a whole (p. 131).*

Brown (1995) argues that ecocentrism may evolve into an ideology of power and control in the same way the anthropocentric ideologies have. This is especially true if individual rights are suppressed in favour of what is good or right for the whole (see Burms 1991, cited in Drenthen 2011, who argues that ecocentrism is an extreme form of anthropocentrism because we ascribe value on the basis of absolutes. 'Value' outside the human perspective, has no meaning). Furthermore, the radical egalitarianism of ecocentrism advocated by deep ecologists and ecofeminists can be just as nihilistic as anthropocentrism, Brown argues, because it manifests itself in the foundation of oppressive ideologies, i.e., that certain oppressive classes, be they capitalist, Marxist, Christian, or male, position themselves to act in the name of all humanity. There is a devaluation of the interests of the weaker classes and a strengthening of the more dominant class(es). Brown continues:

> *By addressing their moral imperative only to the human community, deep ecologists are implicitly recognizing that human beings have a unique place in the cosmos; human beings alone are capable of good and evil. When the deep ecologists ask us to recognize inherent values in nature, they explicitly demand that we refrain from seeing humanity as the sole source and creators of value but they implicitly see humanity as standing alone in possessing the power to recognize value beyond the immediate interests of our species (p. 200).*

The second main criticism of ecocentrism, environmental fascism, corresponds to a deliberate subordination of the well-being of all creatures according to the needs of the broader community (Regan 2004). Individuals may be sacrificed for the greater good of the biotic community, according to Regan (and Brown 1995, above), in maintaining the stability and integrity of the natural world. Regan argues that it is difficult to see how the rights of the individual could find a place in the land ethic scheme. Regan (2004) continues:

> *If, to take an extreme, fanciful but, it is hoped, not unfair example, the situation we faced was either to kill a rare wildflower or a (plentiful) human being, and if the wildflower, as a 'team member,' would contribute more to 'the integrity, stability, and beauty of the biotic community' than the human, then presumably we would not be doing wrong if we killed the human and saved the wildflower. The rights view cannot abide this position, not because the rights view categorically denies that inanimate objects can have rights...but because it denies the propriety of deciding what should be done to individuals who have rights by appeal to aggregative considerations, including, therefore, computations about what will or will not maximally 'contribute to the integrity, stability, and beauty of the biotic community.' Individual rights are not to be outweighed by such*

> *considerations (which is not to say that they are never to be outweighed). Environmental fascism and the rights view are like oil and water: they don't mix (p. 362).*

Collections of systems like forests might have inherent value, but Regan (2004) finds it difficult to understand how moral rights are attributed to these collections of trees or ecosystems. These are not individuals and, as such, are not meriting of moral status (see Steverson 1994). The difference between moral standing and intrinsic value is important to the discussion on environmental ethics. O'Neil (2000) writes that non-sentient environmental entities like rocks and trees may have intrinsic value, but they don't have moral standing. Moral standing is extended to those beings that can be benefited or harmed—if something works for or against their interests. A sentient being, like a dog, can have moral standing and intrinsic value. By contrast, a beautiful picture may have intrinsic value if it has a good outside of the pleasure it may bring, but it is inconceivable to argue that it has moral standing in as much as its interests can be benefitted or harmed.

These concerns have prompted animal rights theorists to suggest that ecocentrism is just another form of utilitarianism where decisions are made for the greatest good of the greatest number. Wildlife has value only if society attaches value to it (Garner 2004). Garner calls this type of anthropocentric environmentalism 'eco-imperialism' because wildlife conservation is more about conflict between competing human interests than about conflict between humans and animals—the light green or shallow anthropocentrism perspective discussed above.

Others argue that species do possess interests that each of their individual members may lack (Johnson 1991). What is good for an individual within the species may or may not be good for the species in general, and vice versa. For example, it is in the interests of a forest ecosystem to have periodic fires, even though fires may not be in the interests of component parts of the ecosystem. Similarly, it may be in the best interests of a population of deer to have predators kill the weaker members of the herd, even though this is not in the best interests of the killed deer. The problem is we can't have it both ways. We give priority to the individual or to the species or ecosystem. It all depends on what we value (Bovenkerketal. 2003).

Differences also exist in practice. Sagoff (1984) illustrates that environmentalist and conservation organizations have customarily been interested in ecological issues over humanitarian issues. For example, the Sierra Club and the Wilderness Society place the principles of ecology and population biology over the principles of individual welfare. The opposite holds true for humanitarian groups. Friends of animal groups and humane societies situate the suffering of animals as their first priority. The differences between ecocentrism (environmentalists) and animal liberation are so marked that Sagoff was forced to conclude that, 'Environmentalists cannot be animal liberationists. Animal liberationists cannot be environmentalists' (p. 42).

Empirical studies serve to verify the differences that exist between the mid-green animal liberation group and the deep-green ecocentrists. Bovenkerk et al. (2003) investigated whether or not it was morally right to rescue lost or ill seals in the Netherlands, for the purpose of reintroducing these animals back into the wild. The authors discovered that sheltering on a large scale tends to work against the grain of the ecological principles espoused over the last couple of decades, i.e., we should not interfere with laws of nature. However, they also discovered that there are few reasons not to shelter, and even better reasons to favour sheltering based on circumstance, including: (1) the gravity or extent of suffering of the animal; (2) the extent to which humans are responsible for suffering; (3) the level of naturalness of the ecosystem; (4) responsibility the actor felt due to their function, such as a park manager; (5) the cost versus benefit of rescue; (6) consequences for the health of the ecosystem; (7) the intrinsic value of animals from the wild, or the prospect of their wildness; (8) whether it be better to leave efforts in the hands of professionals over those of the public; and (9) the educational function of the shelter (p. 17).

In a study of the values held by environmental management professionals, Hull et al. (2003) found that not one of the interviewees (despite having a rather vocal animal rights contingency as part of the pool of interviewees), argued explicitly in favour of extending rights to individual organisms. By contrast, interviewees were much more likely to embrace principles of ecocentrism in reference to what we should value and why, i.e., it is good or desirable to protect nature, even if these actions didn't directly impact the agent Even more than this, anthropocentric values dominated peoples' reasoning behind their views on environmental quality. Hull et al. (2003) write that environmental managers often expressed values tied to the importance of income generation by timber and the recreation benefits that people get from natural areas. Furthermore, forestry and wildlife and resource managers, i.e., those that advocate for the wise use of resources in achieving sustainable yields in the vein of the ecocentric mode of thinking, are much more likely to advocate for animal welfare over animal rights (Schmidt 1990).

It is worth noting that some scholars have attempted to find common ground between the ecocentric and animal liberationist perspectives (see for example, Callicott 1980). However, in a subsequent publication, Callicott was forced to concede that there has been 'an increasingly acrimonious divorce between individualistic animal welfare ethics and holistic ecocentric ethics' (Callicott 1988:163). Despite this divorce, Callicott has attempted to re-marry the two streams into a biosocial moral theory. While his approach extends the boundaries of the common middle ground, those steadfast proponents of both ends will see little overall unity. More recently, Morris and Thornhill (2006) have attempted unity through the use of Roger Scruton's views on intrinsic value using the concept of piety or *pietas*. Pietas is defined as, 'The disposition to acknowledge the sacredness and untouchability of things, places, people and customs...because you have no right to encroach upon them, and because they are intrinsically worthy of respect' (Scruton 2000:183–184, as cited in Morris and Thornhill 2006:370). This takes into consideration not only individual lives, but also biodiversity (species and ecosystems) in general.

Ecocentrism and Tourism

Scholars in tourism studies have utilized ecocentric theory in a number of different ways. Examples include, residents scoring higher on ecocentric scales rejecting plans for the development of second homes (Kaltenborn et al. 2008); the disconnect between pro-environmental attitudes and pro-environmental behaviour amongst ecotourists (Jackson 2007); the contrast between technocentrism and ecocentrism (Acott et al. 1998); and the discussion of ecocentrism as one of three main theoretical perspectives that will help realize a process of the decommodification of tourism through the contribution of non-governmental organizations (Wearing et al. 2005; see also Butcher's 2006 anthropocentric response).

What has not been a focus in tourism until quite recently, however, is the use of ecocentrism (or indeed other environmental ethics theories) as it corresponds to the use of animals in tourism (Fennell 2012b). Any use of an animal for tourism purposes that has an effect on that species and its role within the ecological community is relevant to the discussion on ecocentrism. When tourism practitioners make orang-utans box in front of tourists at a Bangkok zoo (Aljazeera, 2006); use captive black market tigers at Thai temples for tourist pleasure (Dell'Amore 2008); fin endangered shark species (McKie 2009); or continue the practice of hare coursing as a blood sport, even though hares are endangered (Bowcott 2010), there are important and broader whole systems debates that are being ignored.

Not unlike the lack of consideration of how ecocentrism could guide our use of animals in practice, there has been a dearth of academic activity investigating how such theory could inform our understanding of tourism, animals and ethics. Examples of the scant work undertaken to date include Fennell and Nowaczek's (2010) use of biocentric theory to examine how interactions with fish may be described as ecotourism or nature-based tourism according to the type of interaction and the level of consumptiveness; and Shani and Pizam's (2008) contrast of three differentethical approaches as these apply to our use of animals in tourism, including environmental ethics, animal welfare, and animal rights (see also Hughes 2001).

More specifically, Kontogeorgopoulos (2009) used ecocentrism as a central theme in his work on elephant camp tourism in Northern Thailand. Those camps categorized as more ecocentric treat elephants with a higher level of regard, and this sentiment is represented in the preferences and values of tourists who visit these camps. However, even though anthropocentric camps focus more on activities that place emphasis on the recreation and entertainment of tourists, these places also have a role to play in helping elephants. Kontogeorgopoulos argues that tourism is the lesser of many other evils— elephants, aside from tourism, 'possess few legal options for earning a living' (p. 445). Because of diminishing habitat, the expense of caring for animals (presumably in sanctuaries), and the difficulty of reintroducing animals into the wild, working in tourism is viewed as the best option. Kontogeorgopoulos states that

> *Critics of the use of elephants in tourism should acknowledge the financial utility and value that anthropocentric camps create for elephants, while camp owners should seek ways of improving their products by balancing the preferences of tourists for anthropocentric activities such as elephant riding with the more naturalistic ethos of ecocentric camps, which, it would appear, is not only good for business and beneficial for the welfare of domesticated elephants, but is also more rewarding and educational for visitors... (p. 445).*

Ecocentric theory has also been used comprehensively in the case of Fraser Island, Australia. The death of a nine-year-old boy from a dingo attack provoked re-evaluation of policies on how best to manage tourists and

Table 1. Seven Principles for an Ecocentric Approach to Wildlife Tourism.

Principle	Description
Intrinsic value	Wildlife tourism has value in and of itself independent of its usefulness to human activities, including tourism.
Moral obligations	Enabling visitors to become aware of the environmental consequences of their actions can compel them to change their behaviour through a sense of moral obligation.
Moral reasoning	Providing information on environmental ethics facilitates visitors in undertaking ecocentric moral reasoning.
Precautionary principle	If a wildlife tourism action has a suspected risk of causing harm to animals or their habitat, in the absence of scientific consensus that the action is harmful, then the burden of proof that it is not harmful falls on those proposing the action.
Avatar	The interconnectedness of humans and nature requires management of both in a more holistic framework as part of shared ecosystems.
Belonging in nature	An imperative is humans acknowledging that non-humans belong in and with nature and that humans are visitors to the habitat.
Reflective manager	Essentiality of managers being able to self-reflect on how their ethical position is constructed and could potentially change over time.

Source: Burns et al. (2011)

dingoes, especially given the fact that dingo behaviour has changed as tourism numbers continue to escalate (Thompson et al. 2003). Burns et al. (2011) argue that the way forward on Fraser Island is to incorporate an ecocentric ethic in the management of dingoes and non-consumptive wildlife tourism from the perspective of seven main principles (Table 1). The implementation of an ecocentric perspective 'does not place the rights of humans above those of non-humans, but instead argues that non-human species have the same rights to live and prosper as humans' (p. 182).

Principle 3, Moral Reasoning, argues for the incorporation of ecocentric theory as a mechanism for better management Yet true to the ecocentric perspective, at least the Leopoldian view, it is acceptable to kill animals in preserving the integrity, stability, and beauty of the biotic community. The question remains whether we can in fact argue for the intrinsic value of dingoes on par with humans— in addition to the assertion that dingoes have the same rights as humans—while at the same time killing them if they have detrimental effects on the human population. If dingoes have the same rights as people, it would be unlawful to kill them under any conditions. Accordingly, discussions on rights and intrinsic value lie at the heart of whether or not dingoes, or indeed other animals caught in the intricate web of human use of the natural world, should live or die.

A related question in this case is that although Fraser Island attracts non-consumptive wildlife tourists (this would include ecotourists of a hard path or deep nature), it is debatable whether such tourists would be willing to visit environments where consumptive methods of management like culling are regularly used to control predator populations. The assumption is that a percentage of ecotourists, however small, might be uncomfortable with the decision to cull dingoes under any conditions. Burns et al. (2011) explain that ecocentrism is a logical alternative to anthropocentric management policies in the case of Fraser Island. However, this alternative exists in the absence of other theoretical perspectives that may be useful in expanding the debate.

Billfishing as Ecotourism: Revisited

The debate between Holland et al. (1998, 2000) and Fennell (2000) on the legitimacy of billfish angling (fishing for marlin and sailfish) as a form of ecotourism has parallels with the preceding discussion on deep green ecocentrism and mid-green environmental ethics. In general, Holland et al. (1998) argued that billfishing qualifies as a form of ecotourism when it is characterized by degrees of the following six characteristics:

1. It occurs in a unique natural resource: open ocean beyond the horizon and with few if any other boats in view;
2. It attracts outdoor enthusiasts who are like ecotourists in terms of education, household income, and are dedicated to their chosen recreational activity;
3. There is an element of resource responsibility, and this usually takes the form of following a code of ethics or various norms, such as lower bag limits (e.g., one fish limit per person or boat), catch-and-release and tagging fish for scientific or conservation research;
4. There is support for resource conservation, which in the case of billfishing, relates to money from the tourism industry used for conservation purposes and angler membership in conservation organizations;
5. Billfish angling contributes more to the economy than other resource uses like commercial fishing; and
6. The billfishing industry provides employment and entrepreneurial opportunities for local people.

Holland et al. (1998) argue that the catch-and- release style of billfish angling in Costa Rica is morally justifiable as ecotourism (over catching and killing), and can be likened to scuba diving, where the non-consumptive nature of the activity sometimes contributes to occasional mortality. The authors conclude by suggesting that there are some important unanswered questions about the mortality of released fish post-catch as this applies to the consumptive and non-consumptive nature of this activity. However, in terms of release and potential mortality rates, these concerns are perhaps more in line with ensuring that the fishing industry is kept viable in the same manner that Ducks Unlimited continues to preserve habitat for more ducks. This observation gains traction when Holland et al. (1998:97–98) comment that

> *In addition to having the private sector infrastructure necessary to attract anglers, there must be a high level of stock abundance to provide for the high quality fishing experience sought by anglers. Billfish angling destinations must compete worldwide; to the extent that abundant fish stocks can be maintained, those regions in close proximity to major angling markets have an advantage.*

Fennell (2000) responded to Holland et al. (1998) by questioning the extent to which billfishing should qualify as a form of ecotourism based on four main factors: intention; animal pain; consumptiveness; and values, behaviour and taxonomy. Billfishing should not be ecotourism based on the intent of the activity, which is to pursue fish as a prime intention with the probable outcome of producing pain and stress. Ecotourism, by contrast; should be more about the avoidance of harm in the process of securing recreational pleasure and satisfaction. Second, and related to the first point, above, is the possibility that fish feel pain (see below) and often the from these encounters. Concern should be raised over the acceptability of a form of ecotourism that promotes this type of human-animal interaction, where the 'occasional mortality of billfish is acceptable akin to the impacts that scuba diving has on reefs. Third, fishing becomes a consumptive activity if fish are killed despite being released. Finally, the values attached to billfishing appear to be more oriented towards sport As such, the adventure tourism label appears to be more appropriate for this activity than 'ecotourism'.

In revisiting this debate so many years later, my purpose is to extend the discussion by employing ecocentric theory in the case of Holland et al. (1998), and mid-green animal liberation theory in the case of Fennell (2000). These are perspectives that perhaps should have been used initially by both sides in strengthening their respective positions.

Where Holland et al. (1998, 2000) fell short was in failing to incorporate ecocentric theory in framing their argument beyond a focus on billfishing as industry, product, economy and experience. Leopold's land ethic perspective is one advocated by wildlife ecologists and other environmental practitioners (and theorists) in validating the priority of whole systems over the interests of individual animals. From this perspective, as long as the ocean environment is not compromised, and as long as billfish stocks are robust enough to accommodate sport fishing, i.e., as long as we act in a way that promotes the good, and not harm, of the marine environment on the whole, actions like fishing are morally praiseworthy. The importance of income generation and other benefits from the use of natural areas is consistent with the views of environmental management professionals, as anthropocentric views tend to dominate decision-making (Hull et al. 2003; Holden 2003). This is the ethics-by-extension platform described by Rowe (1994), whereby natural elements take a back seat to human needs. In these cases, occasional mortality of animals is undoubtedly acceptable as long as the whole system is healthy and as long as human interests are given priority.

In their subsequent response to Fennell (2000), Holland et al. (2000) observe that 'Whether or not fish feel pain falls under the subject of animal rights and care... we see more of the author's personal values here than research results' (Holland et al. 2000:347). The authors demonstrate a lack of care in addressing the broader literature. First, the issue of pain in fish and other animals is not just a matter of animal rights, but rather one that is much deeper. Pain, and here we ought to source the literature on sentience, appeals not only to animal rights, but also utilitarianism as well as animal welfare (Arlinghaus et al. 2007; Huntingford et al. 2007). Second, there is a wealth of data pointing to the fact that fish do in fact experience pain (see Fennell and Nowaczek 2010 for a summary of some of this literature). Added to this is the finding that catch- and-release fishing is more consumptive than generally thought Domeier et al. (2003) found that 26.2% of striped marlin died within five days of release. Injury from catch was a clear predictor of mortality, e.g., all marlin bleeding from the gills ended up dying.

Furthermore, in reference to their point that, 'anglers are interested in the experience of pursuing and catching large billfish rather than necessarily harvesting them' (p. 98), there is no discussion on the issue of whether fishing for pleasure is less morally justifiable than fishing for sustenance (see Balon 2000; Fennell and Nowaczek 2010).

Fennell (2000) fell short due to his support of a Leopoldian ecocentric perspective, where humans 'are part of a larger ecological system, which must be protected and accepted on its own terms' (p. 342). The mid-green ethic would have been more appropriate for Fennell (2000) to apply in this situation given the individualistic nature of animal liberation, i.e., a focus on the value of individuals, not the sacrifice of individuals for the good of the ecological community. In the case of billfishing, the land ethic as representative of the ecocentric or deep green perspective does not adequately protect the interests of individual billfish. Billfish are not trees or ecosystems, but rather sentient beings that can suffer. And we should not let the many marginal human interests (recreational fishing) outweigh the interests that these animals have in not suffering (Singer 2009).

More recently, Fennell (2013) has argued that the divide between consumptive and non-consumptive activities as ecotourism has been fuzzy in the absence of philosophically grounded arguments that otherwise might allow for generative progression. In this regard, Fennell claims that ecotourism practitioners should act according to first principle when it comes to the use of animals for ecotourism purposes. This principle places the well-being of animals over the recreational and commercial interests of humans—interests that are often embedded in the ecocentric ethic, as noted above. Hence, ecotourism practitioners should

> *Reject as ecotourism all practices that are based on or support animal capture and confinement, or other forms of animal use that cause suffering, for human pleasure and entertainment.*
>
> *Embrace as ecotourism interactions that place the interests of animals over the interests of humans. This would include encounters with free-living animals that would have the liberty to engage or terminate interactions independent of human influence*
>
> *(Fennell, In press-b).*

In view of the stress, pain, and consumptive (death of released fish) aspects of billfish angling, the foregoing principle would provide clear guidance over what would pass as the acceptable or unacceptable use of animals drawn into the ecotourism industry. As such, it is clear that there is room for the presence of a reductionist stance based on the interests of individual animals first over the 'integrity, stability, and beauty of the biotic community' maxim of Leopold. The question remains whether ecotourism practitioners have the will to eliminate activities that involve the pursuit, capture, injury, stress, or death of animals in satisfying the recreational interests of ecotourists specifically and general tourists more broadly. These issues require further investigation as they relate to (1) different types of ecotourism (hard vs. soft), (2) nature-based tourism as compared to ecotourism, and (3) ecotourism in comparison to other forms of tourism.

Conclusion

This paper used Curry's (2011) categorization of shallow, mid, and deep green ethics in an effort to more fully understand the depth and range of environmental ethics theory. Emphasis was placed on ecocentrism, especially how it compared to mid-green theories of animal liberation. Knowledge of these different theoretical perspectives was deemed essential in efforts to more closely analyze the many vexing issues tied to billfishing as a form of ecotourism. Previous efforts by Holland et al. (1998, 2000) and Fennell (2000) were largely atheoretical in their attempts to support one side of the debate over the other.

Those who support billfishing as a form of ecotourism adopt reasoning that corresponds to the ecocentric perspective, i.e., support for billfishing as ecotourism comes from placing value on whole systems as a first priority. As long as the integrity, stability, and beauty of the biotic community are preserved, actions are deemed morally praiseworthy. By contrast, those who adopt an animal liberationist approach argue that there are not good reasons to sacrifice individual animals for the betterment of the whole—and this includes fishing for recreational purposes. Contrary to the position held by Holland et al. (1998), billfishing would not qualify as a form of ecotourism because it does not protect the interests of animals as a first priority. On this account, any activity that creates stress, pain, or causes death in animals should not be viewed as ecotourism, because ecotourism as a practice ought to defend these individual interests. The land ethic does not support this view.

This research reinforces Sagoff's (1984) contention that environmentalists and animal liberationists are at odds over such questions, and these differences provide an explanation as to why Holland et al. (1998, 2000) and Fennell (2000) struggled to find common ground over their respective interpretations of ecotourism. I conclude by suggesting that: (1) there is no set philosophical or empirical basis from which to suggest that the prefix 'eco' ought to refer specifically to ecocentrism as the guiding environmental ethic for ecotourism, over other theories like animal liberation; and (2) future studies in this area should intersect theory from environmental ethics with ecotourism, more broadly, in efforts to build an industry that focuses not only on nature preservation, but also on the nature of moral standing—on many levels.

References

ACOTT, T.G., LATROBE, H.L. and HOWARD, S.H. (1998). An Evaluation of Deep Ecotourism and Shallow Ecotourism. *Journal of Sustainable Tourism* 6(3): 238–253.

ADAMS, C.J. (1993). Introduction. In Adams, C.J. (Ed) *Ecofeminism and the Sacred.* New York, NY. The Continuum Publishing Company: 1–9.

ALJAZEERA (2006). Blow to Thailand Monkey Business. Available at http://www.aljazeera.com/archive/2006/04/200841092626122565.html – Accessed on 8 November 2012.

ARLINGHAUS, R., COOKE, S.J., SCHWAB, A. and COWX, I.G. (2007). Fish Welfare: A Challenge to the Feelings-based Approach, with Implications for Recreational Fishing. Fish *and Fisheries* 8(1): 57–71.

BALON, E.K. (2000). Defending Fishes Against Recreational Fishing: An Old Problem to be Solved in the New Millennium. *Environmental Biology of Fishes* 57(1): 1–8.

BIRO, A. (2002). Towards a Denaturalized Ecological Politics. *Polity* 35(2): 195–212.

BOOKCHIN, M. (1990). *The Philosophy of Social Ecology: Essays on Dialectical Naturalism.* New York. Black Rose Books.

BOVENKERK, B., STAFLEU, F., TRAMPER, R., VORSTENBOSCH, J. and BROM, F.W.A. (2003). To Act or Not to Act? Sheltering Animals from the Wild: A Pluralistic Account of a Conflict Between Animal and Environmental Ethics. *Ethics, Place and Environment* 6(1): 13–26.

BOWCOTT, O. (2010). Northern Ireland Bans Hare Coursing. Available at http://www.guardian.co.uk/uk/2010/jun/23/northern-ireland-bans-hare-coursing – Accessed on 9 November 2012.

BROWN, C.S. (1995). Anthropocentrism and Ecocentrism: The Quest for a New Worldview. *Midwest Quarterly* 36(2): 191–202.

BURMS, A. (1991). Antropocentrisme En Ecocentrisme. In Liegeos et al., (Eds) *Aspecten Van Een Christelijke Sociale Ethiek, Colloquium Over Christelijke Sociale Ethiek Naar Aanleiding Van 100 Jaar Rerum Novarum.* Leuven. Bibliotheek van de Faculteit der Godgeleerdheid: 140–144.

BURNS, G.L., MACBETH, J. and MOORE, S. (2011). Should Dingoes Die? Principles for Engaging Ecocentric Ethics in Wildlife Tourism Management. *Journal* of *Ecotourism* 10(3): 179–196.

BUTCHER, J. (2006). *A Response to Building a Decommodified Research Paradigm in Tourism: The Contribution of NGOs* by Stephen Wearing, Matthew McDonald and Jess Ponting, published in Journal of Sustainable Tourism, Vol. 13, No. 5, 2005: 424–455. *Journal of Sustainable Tourism* 14(3): 307–310.

CALLICOTT, J.B. (1980). Animal Liberation: A Triangular Affair. *Environmental Ethics 2:* 311–338.

CALLICOTT, J.B. (1988). Animal Liberation and Environmental Ethics: Back Together Again. *Between the Species* 4: 163–169.

CALLICOTT, J.B. (1989). *In Defense of the Land Ethic.* Albany. State University of New York.

CARSON, R. (1962). *Silent Spring.* Boston. Houghton Mifflin.

CARTER, A. (2005). Inegalitarian Biocentric Consequentialism, the Minimax Implication and Multidimensional Value Theory: A Brief Proposal for a New Direction in Environmental Ethics. *Utilitas* 17(1): 62–84.

CURRY, P. (2011). *Ecological Ethics: An Introduction* (Second Edition). Cambridge, UK. Polity Press.

DELL' AMORE, C. (2008). Black Market Tigers Linked to Thai Temple, Report Says. Available at http://news.nationalgeographic.com/newB/2008/06/080620-tiger-temple.html – Accessed on 8 November 2012.

DESJARDINS, J.R. (2013). *Environmental Ethics: An Introduction to Environmental Philosophy* (Fifth Edition). Belmont, CA. Wadsworth.

DIZEREGA, G. (1992). Social Ecology, Deep Ecology, and Liberalism. *Critical Review* 6(2–3): 305–370.

DIZEREGA, G. (1995). Empathy, Society, Nature, and the Relational Self: Deep Ecology and Liberal Modernity. *Social Theory and Practice* 21(2): 239–269.

DOMEIER, M.L., DEWAR, H. and NASBY-LUCAS, N. (2003). Mortality Rate of Striped Marlin (*Tetrapturus audax*) Caught with Recreational Tackle. *Marine and Freshwater Research* 54: 435–445.

DONOVAN, J. (1990). Animal Rights and Feminist Theory. *Journal of Women and Culture in Society* 15(2): 350–368.

DRENTHEN, M. (2011). Ecocentrism as Anthropocentrism. *Ethics, Policy and Environment* 14(2): 151–154.

ECKERSLEY, R. (1992). *Environmentalism and Political Theory: Toward an Ecocentric Approach.* Albany. State University of New York.

EVANOFF, R. (2007). Bioregionalism and Cross-cultural Dialogue on a Land Ethic. *Edites, Place and Environment* 10(2): 141–156.

FEIBLEMAN, J.K. (1954). Theory of Integrative Levels. *British Journal for the Philosophy of Science* 5: 59–66.

FENNELL, D.A. (2000). Ecotourism on Trial – The Case of Billfish Angling as Ecotourism. *Journal of Sustainable Tourism* 8(4): 341–345.

FENNELL, D.A. (2012a). Tourism and Animal Rights. Tourism *Recreation Research* 37(2): 157–166.

FENNELL, D.A. (2012b). *Tourism and Animal Ethics.* London. Routledge.

FENNELL, D.A. (2012c). Tourism, Animals and Ethics: Utilitarianism. *Tourism Recreation Research* 37(3): 239-251.

FENNELL, D.A. (2013). Contesting the Zoo as a Setting *for* Ecotourism, and the Design of a First Principle. *Journal of Ecotourism* 12(1): 1-14.

FENNELL, D.A. and NOWACZEK, A. (2010). Moral and Empirical Dimensions of Human-Animal Interactions in Ecotourism: Deepening an Otherwise Shallow Pool of Debate. *Journal of Ecotourism* 9(3): 178–189.

FENNELL, D.A. and SHEPPARD, V.A. (2011). Another legacy for Canada's 2010 Olympic and Parlympic Winter Games: Applying an Ethical Lens to the Post-games' Sled Dog Cull. *Journal of Ecotourism* 10(3): 197–113.

FOX, W. (1989). The Deep Ecology-Ecofeminism Debate and its Parallels. *Environmental Ethics* 11(1): 5–25.

FRENCH, W.C. (1995). Against Biospheric Egalitarianism. *Environmental Ethics* 17: 39–57.

GARNER, R. (2004) Animals, *Politics and Morality.* New York. Manchester University Press.

GRUEN, L. (2010). The Moral Status of Animals. In Zalta, E. N. (Ed) *The Stanford Encyclopedia of Philosophy.* Available at http://plato.stanford.edu/entries/maral-animal/ – Accessed on 7 September 2011.

HARDIN, G. (1968). The Tragedy of the Commons. *Science* 162: 1243–1248.

HETTINGER, N. and THROOP, B. (1999). Refocusing Ecocentrism: De-emphasising Stability and Defending Wildness. *Environmental Ethics* 21: 3–21.

HOLDEN, A. (2003). In Need of a New Environmental Ethics for Tourism? *Annals of Tourism Research* 30(1): 95–108.

HOLLAND, S.M., DITTON, R.B. and GRAEFE, A.R. (2000). A Response to "Ecotourism on Trial: The Case of Billfish Angling as Ecotourism'. *Journal of Sustainable Tourism* 8(4): 346–351.

HOLLAND, S.M., DITTON, R.B. and GRAEFE, A.R. (1998). An Ecotourism Perspective in Billfish Fisheries. *Journal of Sustainable Tourism* 6(2): 97–116.

HUGHES, P. (2001). Animals, Values and Tourism – Structural Shifts in UK Dolphin Tourism Provision. Tourism *Management* 22: 321–329.

HULL, R.B., RICHERT, D, SEEKAMP, E., ROBERTSON, D. and BUHYOFF, G.J. (2003). Understandings of Environmental Quality: Vmbiguities and Values Held by Environmental Professionals. *Environmental Management* 31(1): 1–13.

HUMPHREY, M. (2000). 'Nature' in Deep Ecology and Social Ecology: Contesting the Core. *Journal of Political Ideologies* 5(2): 247–268.

HUNTINGFORD, F., ADAMS, C., BRAITHWAITE, V.A., KADRI, S., POTTINGER, T.G., SANDOE, P. and TURNBULL, J.F. (2007). The Implications of Feelings-based Approach to Fish Welfare: A Reply to Arlinghaus et al. *Fish and Fisheries* 8(3): 277–280.

JACKSON, S. (2007). Attitudes Towards the Environment and Ecotourism of Stakeholders in the UK Tourism Industry with Particular Reference to Ornithological Tour Operators. *Journal of Ecotourism* 6(1): 34–66.

JOHNSON, L. (1991). *A Morally Deep World.* Cambridge. Cambridge University Press.

KALTENBORN, B.P., ANDERSON, O. and NELLEMANN, C. (2008). Resident Attitudes Towards Mountain Second-home Tourism Development in Norway: The Effects of Environmental Attitudes. *Journal of Sustainable Tourism* 16(6): 664–680.

KONTOGEORGOPOULOS, N. (2009). Wildlife Tourism in Semi-captive Settings: A Case Study of Elephant Camps in Northern Thailand. *Current Issues in Tourism* 12(5&6): 429—449.

LEHMAN, S. (1981). Do Wildernesses Have Rights? *Environmental Ethics* 3: 129–146.

LEOPOLD, A. (1933a). Game *Management.* New York. MacMillan.

LEOPOLD, A. (1933b). The Conservation Ethic. *Journal of Forestry* 31: 634-641.

LEOPOLD, A. (1970). A *Sand County Almanac with Essays on Conservation from Round River.* New York. Ballantine Books.

LEOPOLD, A. (2004). Living with the Land Ethic. *BioScience* 54(2): 149–154.

LOFTIN, R.W. (1985). The Medical Treatment of Wild Animals. *Environmental Ethics* 7: 231–239.

LOVELOCK, J.E. (1979). Gaia: *A New Look at Life on Earth.* Oxford. Oxford University Press.

MCKIE, R. (2009). Shark Fin Ban Ends Cruel Slaughter. Available at http://www.guardian.co.uk/environment/2009/oct/11/uk-shark-finning-ban-extinct – Accessed on 8 November 2012.

MEHMETOGLU, M. (2009). Predictors of Sustainable Consumption in a Tourism Context: A CHAID Approach. *Advances in Hospitality and Leisure* 5: 3–23.

MORRIS, M.C. and THORNHILL, R.H. (2006). Animal Liberationist Responses to Non-anthropogenic Animal Suffering. *Worldviews* 10(3): 355–379.

NAESS, A. (1984). A Defence of the Deep Ecology Movement. *Environmental Ethics* 6(3): 265–270.

NASH, R.F. (1989). *The Rights of Nature: A History of Environmental Ethics.* Madison. The University of Wisconsin Press.

NELSON, M.P. (2004). Teaching the Land Ethic. *Worldviews* 8(2&3): 353–365.

NORTON, B.G. (1984). Environmental Ethics and Weak Anthropocentrism. *Environmental Ethics* 6(2): 131–148.

O'NEIL, R. (2000). Animal Liberation Versus Environmentalism: The Care Solution. *Environmental Ethics* 22: 183–190.

OUDERKIRK, W. (2000). "Forever Wild" and Wild Philosophy: The Adirondacks and Environmental Ethics. *Adirondack Journal of Environmental Studies* (Fall/Winter): 23–29.

PAYNE, P.G. (2010). The Globally Great Moral Challenge: Ecocentric Democracy, Values, Morals and Meaning. *Environmental Education Research* 16(1): 153–171.

REGAN, T. (2004). *The Case for Animal Rights.* Berkeley. University of California Press.

ROLSTON III, H. (1981). Values in Nature. *Environmental Ethics* 3: 113–128.

ROLSTON III, H. (1975). Is There an Ecological Ethic? Ethics: *An International Journal of Social, Political, and Legal Philosophy* 18(2): 93–109.

ROWE, S. (1994). Ecocenrtrism: The Responsive Chord. *Trumpeter* 11(2). Available at http://trumpeter.athabascau.ca/index.php/trumpet/article/viewArticle/330/508 – Accessed on 4 November 2012.

ROWE, J.S. (1997). From Reductionism to Holism in Ecology and Deep Ecology. *The Ecologist* 27(4): 147-151.

SAGOFF, M. (1984). Animal Liberation and Environmental Ethics: Bad Marriage, Quick Divorce. *Osgoode Hall Law Journal* 22: 297–307.

SCHMIDT, R.H. (1990). Why Do We Debate Animal Rights? *Wildlife Society Bulletin* 18: 459–461.

SCRUTON, R. (2000). *Animal Rights and Wrongs* (Third Edition). London. Metro Books.

SESSIONS, G. (1995). Ecocentrism and the Anthropocentric De-tour. In Sessions, G. (Ed) *Deep Ecology for the Twenty-First Century.* Boston. Shambala: 156–183.

SHANI, A. and PIZAM, A. (2008). Towards an Ethical Framework for Animal-based Attractions. *International Journal of Contemporary Hospitality Management* 20(6): 679–693.

SINGER, P. (2009). *Animal Liberation: The Definitive Classic of the Animal Movement.* Toronto. Harper Perennial.

STEVERSON, B.K. (1994). Ecocentrism and Ecological Modelling. *Environmental Ethics* 16: 71–88.

TAYLOR, P. (1986). *Respect for Nature.* Princeton. Princeton University Press.

THOMPSON, J., SHIRREFFS, L. and MCPHAIL, L (2003). Dingoes on Fraser Island—Tourism Dream or Management Nightmare. *Human Dimensions of Wildlife* 8: 37–47.

WEARING, S., MCDONALD, M. and PONTING, J. (2005). Building a Decommodified Research Paradigm in Tourism: The Contribution of NGOs. *Journal of Sustainable Tourism* 13(5): 424–455.

WHITE, Jr, L.T. (1967). The Historical Roots of Our Ecologic Crisis. Science 155 (3767): 1203–1207.

YUDINA, O. and FENNELL, D.A. (2013). Ecofeminism in the Tourism Context: A Discussion of the Use of Other-than-human Animals as Food in Tourism. *Tourism Recreation Research* 38(1): 55–70.

Submitted: November 30, 2012
Accepted: February 24, 2013

Sustainability and ecotourism principles adoption by leading ecolodges: learning from best practices

Danuta de Grosbois and David A. Fennell

ABSTRACT

This study identified best practices in the sustainable management of leading ecolodges around the world based on a qualitative content analysis of sustainability communications. Sixty-five top global ecolodges were identified and analyzed according to (1) independent third party ecolodge designations, and (2) a conceptual framework of ecolodges based on a rigorous definition of ecotourism, emphasizing a nature-based focus, education, sustainability (conservation and community involvement/benefits), and moral imperative dimensions. Key themes within these dimensions included biodiversity conservation, preserving socio-cultural heritage, improving social wellbeing, contributing to the local economy, learning opportunities for guests, influencing local attitudes, creating opportunities for visitors to get involved, sharing profits with local people, social justice, and seeking third-party recognition. This study contributes to the expanding base of literature on ecolodges by investigating best practices within each core dimension of ecotourism and evaluating these through corporate social responsibility and stakeholder theory perspectives.

Introduction

Ecotourism destinations around the world increasingly view ecotourism as an attractive way to grow tourism, given its potential to generate economic benefits without harming the environment. It is perceived as a useful tool for biodiversity conservation and ecological restoration (Blangy & Mehta, 2006; Epler Wood, 2002), education of visitors about sustainability, protection of local livelihoods and culture, empowerment of local communities (Scheyvens, 1999) and a source of learning opportunities, personal growth and heightened self-esteem for local community members (Stronza & Gordillo, 2008). However, in order to generate these benefits, the implementation of ecotourism needs careful consideration, including proper policies, adequate site planning and development of necessary infrastructure, ranging from accommodation facilities to trails, campsites and associated support systems. The construction and operation of ecolodges is a crucial factor determining the success of ecotourism as they represent an essential component of the ecotourism experience.

Since ecolodges are built in natural areas where ecotourism experiences take place, it is expected that they meet the philosophy and principles of ecotourism (Mehta, 2007). Since the 1990s, several guidelines have been developed to assist ecotourism operators and developers in better understanding ecolodge design and operations. Some of these guidelines explicitly focus on ecolodges (Hawkins et al., 1995; International Finance Corporation, 2004; Mehta et al., 2002; Sanders & Halpenny, 2001), while others address ecotourism and ecolodges in general (Drumm et al., 2004; TIES, 1993; Wood, 2002). According to Drumm et al. (2004), the development of ecolodges should follow the principles of sustainability, including human design with the natural world, relying on natural energy flows, eliminating waste and balancing 'human needs with the capacity of the natural and cultural environments' (p. 29).

However, the implementation of ecotourism principles and sustainability can pose significant challenges for ecolodge operators, and there is growing concern that not all lodges are sustainable (Mehta, 2007). A better understanding of factors influencing the adoption of different practices and what best practices are would provide valuable guidance for operators. Despite the importance of ecolodges, empirical research on ecolodges is limited (Kwan et al., 2010). Ecolodges are included as part of several studies on the implementation of sustainability practices by eco-accommodation (Nelson, 2010) or by ecotourism operators (Donohoe &

Needham, 2008; Massi & De Nisco, 2018). Studies devoted exclusively to ecolodges focus on tourists' experiences and satisfaction (Chan & Baum, 2007; Lu & Stepchenkova, 2015; Mafi et al., 2019) or ecolodges' performance evaluation (Osland & Mackoy, 2004). Several studies evaluated ecolodges' adoption of ecotourism principles (Lai & Shafer, 2005; Millar & Sammons, 2006; National Geographic, 2019; Sangpikul, 2010; Wallace & Pierce, 1996). However, these studies do not investigate best practices within each dimension of ecotourism, so there is limited understanding of how these principles translate into best practices in the sustainable operation of ecolodges.

Identification of best practices in sustainability management can serve as a benchmark on how ecolodge operations can be improved. Bretschneider et al. (2005) argue that best practices have three essential characteristics: a comparative process, since it is implied that they are better compared to other possible courses of action; the action itself; and a linkage between the action and some outcome or goal it is designed to achieve. Best practices, therefore, are those that lead to the best performance. However, since the identification of the link between an action and the outcome can be challenging, Laugen et al. (2005) proposed to define best practice as the practices implemented by the best-performing companies. Therefore, in order to identify best practices, the best performers and their practices should be studied. Following this approach, the purpose of this study is to identify best practices in the sustainable management of the ecolodge sector. To meet this purpose, this study develops a conceptual model of ecolodge sustainability and conducts a qualitative content analysis of sustainability communications available on websites of leading ecolodges around the world.

Ecolodges and ecotourism principles

According to Wight (1997), ecotourism accommodation ranges from non-fixed roof units like tents and campsites, to fixed roof accommodation such as cabins and lodges located either in the resource base where ecotourism activities and experiences take place or in nearby villages or cities. The term ecolodge was introduced by the International Ecotourism Society (TIES) during its first International Ecolodge Forum held in 1994 in the US Virgin Islands to explicitly identify and distinguish lodges that are nature-depended and meet the philosophy and principles of ecotourism (Mehta, 2007). Since then, more ecolodge definitions have surfaced to diversify our understanding of the concept. Wallace and Pierce (1996) listed six principles that an ecolodge should follow: minimize negative social and ecological impacts, raise visitors' awareness of natural and cultural systems, contribute to the conservation of protected areas, ensure participation of local people, share economic with local people without overwhelming or replacing traditional practices, and create opportunities for local people to visit natural areas. In a more recent definition, Mehta (2007) specified that 'an ecolodge is a five- to 75-room, low-impact, nature-based, financially sustainable accommodation facility that helps protect sensitive neighbouring areas; involves and benefits local communities; offers tourists an interpretive and interactive participatory experience; provides spiritual communion with nature and culture; and is planned, designed, constructed and operated in an environmentally and socially sensitive manner' (Mehta, 2007, p. 419). Osland and Mackoy (2004) further distinguished between dedicated, casual, scientific and agri-ecolodges based on lodge setting and type of activities offered to visitors. As ecolodges need to satisfy the main principles of ecotourism (Mehta, 2007), the correct definition of ecotourism is essential. Definitions provide the necessary foundation from which policies and practices are actuated: good definitions lead to good policies, which in turn lead to good practices. Failing to properly conceptually and theoretically situate moral enterprises like ecotourism (Nowaczek & Smale, 2010), opens the door for greenwashing and the inclusion of other activities that may not represent the spirit of ecotourism, such as hunting (Fennell, 2020).

Ecolodge goals, sustainability practices and adoption of ecotourism principles have been addressed in several previous studies (Lai & Shafer, 2005; Millar & Sammons, 2006; National Geographic, 2019; Osland & Mackoy, 2004; Wallace & Pierce, 1996). Wallace and Pierce (1996) evaluated eight ecolodges in Amazonas against six ecotourism principles using observation and interviews with visitors, locals and employees and concluded that the ecolodges only partially satisfy the set of ecotourism principles. In a study of 21 ecolodges in Costa Rica and Mexico, Osland and Mackoy (2004) identified 84 performance goals used by owners and managers. The objectives most frequently cited by respondents included profit, preservation of the forest, benefit the local community and educate tourists. The authors applied two-factor theory (Herzberg et al., 1959) to highlight the importance of meeting financial goals as a necessary condition for a lodge to survive, but not sufficient to consider it successful. Lai and Shafer (2005) focused on Internet communication of 35 ecolodges in Latin America and the Caribbean and similarly to Wallace and Pierce (1996) concluded that most of them only partially aligned with ecotourism principles. Weaver and Lawton (2002) developed a typology of ecolodge visitors to Lamington National Park in Australia. Ecolodge visitors were found not to be a homogeneous group. They could be arranged into one of three clusters: 'Harder' ecotourists were characterized as seeking more substantial environmental commitment, enhancive sustainability, specialized longer and active trips, smaller groups sizes, with fewer services expected and an emphasis on personal experience. 'Softer' and 'Structured' ecolodge participants had less of a focus on these characteristics, respectively. The heterogeneity of ecolodge patrons regarding trip characteristics and motivations has been shown in other studies (see the work of Kwan et al., 2008 on ecolodges in Belize).

Despite the limited amount of research on ecolodge sustainability, valuable insight can be gained from the literature on ecotourism and corporate social responsibility (CSR), which focuses on business responsibilities towards the environment and society. As customers increasingly expect companies to embed sustainability values in their products, services and actions (Lii & Lee, 2012), interest in CSR has also been growing both in academia and practice, and CSR has become the subject of several comprehensive literature reviews (Aguinis & Glavas, 2012; Coles et al., 2013). One of the more widely used definitions of CSR states that it 'encompasses the economic, legal, ethical, and discretionary expectations that society has of organizations at a given point in time' (Carroll, 1979, p. 500). It is often conceptualized as the 'triple-bottom-line' approach where companies, in addition to economic goals, also voluntarily pursue social and environmental goals (O'Rourke, 2003). There is now evidence pointing to the fact that the ecolodge itself can be a source of satisfaction through environmental management and education. For example, Hakim and Nakagoshi (2014) investigated the connection between climate change and sustainability in ecolodges through efforts in the area of climate change mitigation standard of responsibility in ecolodge operations.

While several management theories have been applied to CSR research, the present study is informed by stakeholder theory. Stakeholder theory emphasizes the obligation and responsibility of a company to address not only its own goals but also the goals and needs of its stakeholders, defined as individuals or groups that can affect or are affected by the organization's decisions (Freeman, 1984). Proponents of this theory argue that organizations need to engage different stakeholders in a collaborative decision-making process which considers the needs of multiple parties. The involvement of stakeholders in business decisions is both the responsible thing to do as well as a beneficial strategy to improve performance (Plaza-Úbeda, 2010). The core hypothesis of instrumental stakeholder theory states that developing highly ethical stakeholder relationships characterized by fairness, trustworthiness, cooperation, care, respect and information sharing will lead to competitive advantage and improved financial performance of the operator (Hendry, 2001; Jones et al., 2018). Further, according to stakeholder theory, firms disclose information about their sustainability efforts and practices in response to stakeholders' demand for information (Deegan, 2002).

In the context of ecolodges, many stakeholders with varying information needs can be identified, including the owners, managers, employees, local communities, suppliers, NGOs, or local governments. The premise of building ethical relationships with stakeholders is especially relevant to the ecolodge context, as it constitutes one of the core ecotourism principles. Furthermore, stakeholder theory provides a conduit by which to understand how ecolodge operations can influence owners and employees through direct management practices, local community through employment and procurement practices, tourists though the choice of activities offered, and the environment through design and ongoing management. Research on small tourism businesses indicates the vital role of owners as stakeholders, given that their values and lifestyle strongly influence the adoption of sustainable practices by the business (Carlsen et al., 2001; Tzschentke et al., 2008). If poorly designed, built or operated, ecolodges have the potential to harm different stakeholder groups, especially local communities and the environments upon which they rely (Millar & Sammons, 2006; Osland & Mackoy, 2004). Osland and Mackoy (2004) pointed out that whether a given lodge could be considered an ecolodge can be challenging and depend on ideology. As there are no internationally accepted guidelines or certifications for ecolodges, there is considerable confusion in the ecolodge market, where nature-based tourist lodges, such as fishing and ski lodges may also be called ecolodges.

Conceptual framework

Figure 1 presents the conceptual framework that guides this study. Following ecotourism and corporate social responsibility streams of research, the framework proposes that ecolodge sustainability and its best practices are shaped by ecotourism principles as well as internal and external enablers and barriers. Fennell (2020) argued that the core essence of ecotourism involves four main dimensions: a nature-based focus; learning

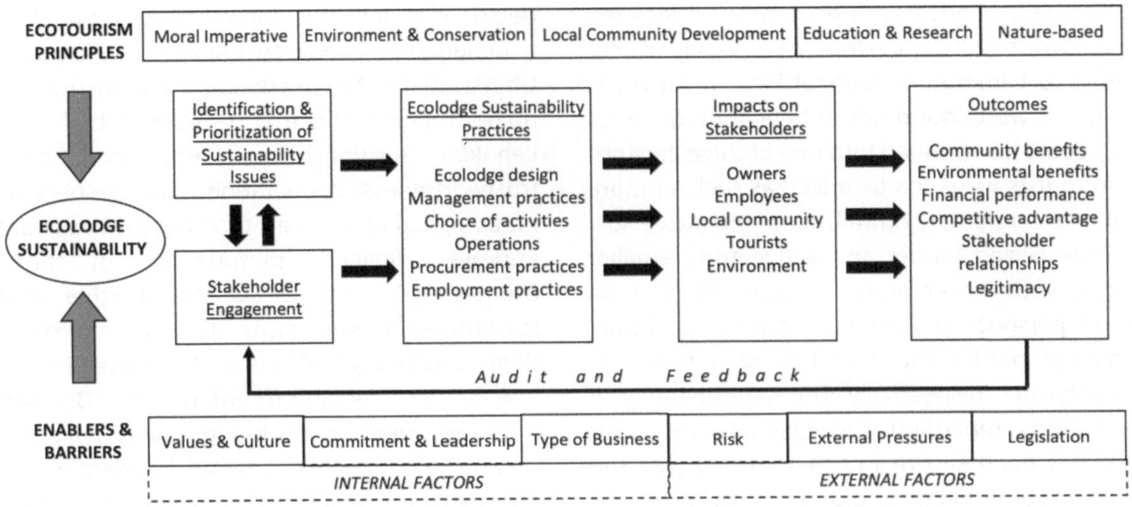

Figure 1. Model of best practices in ecolodge sustainability.

and research; sustainability from the perspective of biodiversity conservation and local involvement and benefits; and a moral imperative. While there is a tacit agreement on the first three of these dimensions, there is mounting evidence to support the notion that ecotourism operations ought to be ethical in their planning, development and management in efforts to separate ecotourism from other nature-based activities like fishing and hunting (Nowaczek & Smale, 2010).

The underlying premise driving the framework in Figure 1 is the view that ecolodges choose their sustainability practices based on the process of identification and prioritization of sustainability and ecotourism issues relevant to their operations. Such issues can include social equity, health and safety, pollution or poverty. The relevance and significance of different issues are subjective, and engagement with stakeholders is a vital part of the process of their identification. External and internal enablers and barriers affect the process and result in different issues. Understanding and prioritization of sustainability issues lead to the implementation of specific, corresponding practices at an ecolodge, such as management, employment, or procurement practices. Implementation of these practices is also affected by internal factors and external factors, such as costs, feasibility, or risk. The choice of practices and their implementation affects the stakeholders and leads to specific outcomes both for the ecolodge and its external environment. The outcomes include social, environmental and financial performance of the ecolodge, reflecting the triple bottom line approach to organizational sustainability. It is essential to recognize the importance of and interactions among all three dimensions. Good financial performance enables the ecolodges to pursue other organizational goals such as supporting the local community, protecting biodiversity or promoting education (Osland & Mackoy, 2004) and is critical to its long-term survival and sustainability (Mehta, 2007). Inability to generate sufficient income to cover expenses would result in a business failure, so it is critical that ecolodges carefully balance their sustainability practices with strategic and tactical management practices to ensure business viability and long-term competitiveness. Sustainability or corporate social responsibility practices can lead to improved performance of ecolodge, in terms of greater customer loyalty and satisfaction, better efficiency, more engaged and motivated employees and better relations with the local community and public authorities (Kim & Park, 2011; Sen & Bhattacharya, 2001). Adoption of CSR practices can further become a form of reputation insurance, generating moral capital and goodwill by enhancing the company's image and reputation among stakeholders, which in turn can serve as insurance protecting the company in the aftermath of adverse events (Godfrey et al., 2009). Customers are also likely to reward companies that engage in CSR activities by purchasing more of their products and services and giving them higher evaluations and positive feedback (Sen & Bhattacharya, 2001). The outcomes of CSR efforts need to be communicated to stakeholders and their feedback incorporated in the process of evaluation and improvement of the practices. The proposed conceptual model further suggests that the sustainability priorities and the related practices implemented by a given ecolodge are continually evolving, in response to internal evaluation as well as external feedback. As a result, different ecolodges can be at different stages of their sustainability system development. The feedback loop in the model illustrates the impact of achieved outcomes and performance on the prioritization of issues and choice of sustainability practices to be implemented.

Methodology

In order to identify organizational best practices, top-performing companies need to be studied (Laugen et al., 2005). Given the purpose of this study to identify best practices in the ecolodge sector, it was necessary to investigate the ecolodges that have been identified as top performers in this sector and could be considered leaders. This was accomplished by adopting a purposive sampling strategy not based on ecolodge self-declarations, but rather on independent third party ecolodge designations that would indicate superior performance. It was expected that these ecolodges would have a high degree of compliance with ecotourism principles and as such would be a suitable sample to study in order to meet the research goal of the paper, which was to identify the best sustainability practices in ecolodge sector. An extensive search of the literature yielded a sample based on the following publications:

(1) lodges listed in the Authentic Ecolodges book (Mehta, 2010)
(2) lodges that were recipients of the World Travel Awards in the ecolodge category during years 2007–2019
(3) lodges listed by the National Geographic on the 50 top ecolodges list (Christ & Siber, 2008)

After removing duplicate ecolodges from these three sources, the resulting sample consisted of 83 ecolodges. Further, in order to ensure the similar nature of operations, only lodges that were permanent and fixed structures were included in the study (see Wight, 1997, above). As a result, 13 tent-based lodges were removed, leading to a sample of 70 ecolodges.

This study used qualitative content analysis (Stemler, 2000) of the written communications on sustainability and ecotourism provided on the ecolodges' websites. During the process of identifying websites for each ecolodge, it was discovered that five ecolodges did not have a website and therefore were removed from the study, resulting in the final sample of 65 properties, listed in Table 1. The majority of the lodges (67.7%) were operated as single properties, with their owners not owning or managing any other lodges, or as single conservation areas, whose owners operated several lodges in the same reserve, but did not own or operate any properties beyond that reserve. The remaining lodges were each part of a larger group of properties or were owned by a larger company such as a tour operator. The lodges were located mostly in Africa (33.8% of the sample), Central and South America (21.5%) and Asia (20.0%). The remaining lodges were in Oceania, North America, the Caribbean, the Middle East and Europe.

The data collected included all content related to ecotourism and sustainability principles, in the form of different documents and information directly displayed in the website's links, but excluding information in the form of videos and links to external articles. If an ecolodge had a link to its parent company website or owner's website, the information on these websites was also included if it applied to the specific ecolodge

Table 1. List of ecolodges included in the final sample.

Ecolodge Name, Location	Ecolodge Name, Location
Adrère Amellal Eco Lodge, Egypt	Kumbuk River, Sri Lanka
Al Maha Desert Resort & Spa, United Arab Emirates	Lapa Rios Lodge, Costa Rica
Apani Dhani Ecolodge, India	Manda Wilderness Nkwichi Lodge, Mozambique
Aristi Mountain Resort & Villas, Greece	Mashpi Lodge, Ecuador
Basata Ecolodge, Egypt	Milia Mountain Retreat, Greece
Bay of Fires Lodge, Australia	Morgan's Rock Hacienda & Ecolodge, Nicaragua
Black Sheep Inn Ecolodge, Ecuador	Mosetiha Bush Camp, South Africa
Borana Lodge, Kenya	NAPO Wildlife Center, Ecuador
Bushmans Kloof Wilderness Reserve & Wellness Retreat, South Africa	Nihi Sumba (formerly Nihiwatu), Indonesia
Caiman Ecological Refuge, Brazil	North Island, Seychelles
Campi ya Kanzi, Kenya	O'Reilly's Rainforest Retreat, Australia
Canopy Tower, Panama	Phinda Forest Lodge, South Africa
Chalalàn Ecolodge, Bolivia	Posada Amazonas Lodge, Peru
Chumbe Island Coral Park, Tanzania	Ranweli Holiday Village, Sri Lanka
Coconut Lagoon, India	Rosalie Forest Eco Lodge, Dominica
Cree Village Eco Lodge, Canada	Sanctuary Olonana, Kenya
Cristalino Jungle Lodge, Brazil	Sarova Shaba Game Lodge, Kenya
Crosswaters Ecolodge & Spa, China	Shamwari Game Reserve – Eagles Crag Lodge, South Africa
Daintree Eco Lodge & Spa, Australia	Six Senses Hideway Ninh Van Bay, Vietnam
Ecolodge Rendez-Vous, Saba Island	Soneva Fushi, Maldives
El Monte Sagrado, New Mexico, USA	Sukau Rainforest Lodge, Malaysia
Feynan Ecolodge, Jordan	The Boat Landing Guest House, Laos
Finca Esperanza Verde Ecolodge, Nicaragua	The Lodge at Chaa Creek, Belize
Highland Center, New Hampshire, USA	Tiamo Resort, Bahamas
Il Ngwesi Eco-Lodge, Kenya	Tiger Mountain Pokhara Lodge, Nepal
Jock Safari Lodge, South Africa	Topas Ecolodge, Vietman
Kapawi Ecolodge, Ecuador	Tourindia Kettuvaliarn Houseboats, India
Kasbah Du Toubkal, Morocco	Tswalu Kalahari Reserve, South Africa
Kaya Mawa, Malawi	Uruyén Lodge, Venezuela
Kicking Horse River Lodge, Canada	Vamizi Island Resort, Mozambique
Kleins Camp, Tanzania	Vumbura Plains Camp, Botswana
Knapdale Ecolodge, New Zealand	Wilderness Lodge Arthur's Pass, New Zealand
	Wolwedans, Namibia

being evaluated. If its foundation ran sustainability-related efforts of a given ecolodge, the information provided on the foundation website was included if it was linked from the ecolodge's website. In order to ensure comparability of the provided information and the potential for the website content to change over time, all the websites were coded within a short period (May to June 2020).

Content analysis has been used previously both in corporate social responsibility (Gray et al., 1995) and ecotourism research (Donohoe & Needham, 2008; Fennell, 2001; Lai & Shafer, 2005). Since the study was qualitative, the data analysis involved organizing data based on themes, concepts, or other similar features through the use of open, axial, and selective coding. Coding is 'the process of grouping evidence and labelling ideas so that they reflect increasingly broader perspectives' (Creswell & Plano Clark, 2007). The coding unit was a paragraph. The data exploration stage involved organizing the text, reading through all of the data in order to develop a preliminary understanding and writing short memos. Qualitative codes were generated based on past literature as well as codes that emerged during the analysis. The initial codes included four ecotourism principles, as proposed in the comprehensive ecotourism framework developed by Fennell (2020) and included in the conceptual model developed for this study. These dimensions are a nature-based focus, moral imperative, sustainability (divided into two sections, including environment and conservation, local community development), and education and research. The content of each website was reviewed in-depth, and the practices contributing to each ecotourism principle were identified. The data was then organized into themes, sub-themes and best practices within each ecotourism dimension. These categories were identified during data analysis through the review and coding process, which involved adding new codes, removing other codes, as well as renaming and merging themes and groups of practices.

Results

The data analysis indicated that there were 20 main themes within the four ecotourism dimensions. Each theme included several best practices that allowed the sector leaders to outperform their competitors concerning ecotourism and sustainability. The following sub-sections discuss the themes and best practices in more detail.

Nature-based focus

Ecolodge practices related to a nature-based focus were grouped into three main themes related to location, facilities and activities (Table 2). Locating ecolodges in protected areas, with high biodiversity and distant from developed areas were the leading best practices related to location. The lodges in the sample were located in rainforests, jungles, temperate forests, savannas, mountains, deserts or water-based locations including islands, bays or rivers. The locations offered access to

Table 2. Nature-based focus best practices adopted by the leading ecolodges.

Themes	Best practices
Location in natural area	• Protected area: within national park, protected sanctuary, within private conservation area or private nature reserve • Access to area with high biodiversity or unique features: providing opportunities for viewing wildlife, e.g. rainforest, savanna, islands close to coral reefs, on banks of important waterways • Remote area: far from developed area, hard to access
Accommodation and infrastructure with good access to nature	• Lodge construction: treehouse-style accommodation high in the rainforest canopy with view of forest and wildlife • Facilities: private ocean-view deck; lodge with view on waterhole; oversized hammocks; beach huts; no television, radio, in-room Wi-Fi or telephones to facilitate connection to nature; relaxation areas to enjoy the scenery, roof terraces • Hiking or biking trail system: including bird stops, camera traps; offering wide variety of hikes with a variety of difficulty levels and duration/length
Nature-related activities	• Nature observation: wildlife viewing and tracking, bird watching, nature study, nature photography (trips with expert photographic guide and a fully-equipped photographic vehicle), searching for rare, endangered or undiscovered animals, tours to observe wilderness scenery, natural landscapes, vegetation and wildlife • Guided nature tours: rainforest interpretation, explanation of ecological processes and habitats, explanations of animal behaviours • Land/air activities: hiking, visiting national parks, climbing, horseback riding, beach activities, biking, stargazing, camping, canopy zip lining, bush drives, air/walking safaris, infrared night safaris • Water activities: kayaking, canoeing, snorkelling, cruises, boat trips along the coastline, rafting, swimming, diving, scuba safaris, intertidal walks during low tide • Dedicated packages: bird lovers (sunset birdwatching cruise, complimentary binoculars, aviary diary), couples (picnicking in nature, sunset or sunrise river cruise), families, adventurers, escapists • Customized experiences: trips to fit visitors' nature-related interests

diverse landscapes, vegetation and wildlife. Secondly, lodges focused on providing accommodation, facilities and infrastructure that supported access and interaction with nature. Best practices within this theme focused on enabling visitors to experience nature's sounds and views from their rooms or areas around the lodge. The final theme encompassed practices related to developing and offering nature-based activities tourists could engage in while visiting the ecolodge. These practices include nature observation, guided nature tours and diverse land, air or water activities. The activities could be grouped in general packages or dedicated nature-based packages aimed at specific market segments, such as families, couples, bird lovers, and so on. Finally, lodges in the sample also offered customized packages to address individual visitors' interests and needs. The number and range of different experiences and packages differed between the lodges. Some offered a wide range of activities, while others specialized in one type, for example, activities related to bird watching.

Sustainability: environment and conservation

The dimension of 'Environment and Conservation' focused on the sample's efforts to protect the natural environment and biodiversity. The relevant best practices within this dimension belong to three different themes: lodge design and construction, environmentally friendly lodge operations and biodiversity conservation (Table 3).

The best practices within the lodge design and construction addressed aspects such as the lodges' location within the natural area, type of materials used in construction with a focus mostly on using local and sustainable materials, integration with nature and energy-efficient design and construction. The practices related to environmentally-friendly ecolodge operations were grouped into sub-themes of water, energy, waste and supply chain management as well as pollution and carbon emissions reduction. Finally, practices focused on biodiversity conservation included sustainable land management, restoration of wildlife, reforestation, collaboration with external organizations, financial contributions, reduction of impacts and support for responders. Different lodges prioritized different issues. For example, when considering location, some lodges stated that they wanted to prevent soil erosion and moved further away from riverbanks, while others were more concerned with animal movements and chose locations that would minimize the impact on animals.

Similarly, concerning energy management, the lodges in the sample ranged from those that eliminated some uses of electrical energy, such as electrical lighting or air conditioning, to those that still provided these uses, but at a reduced scale, e.g. limiting the number of hours air conditioners operate or by using solar-powered lamps. Some of the identified practices have broad applicability and were used by many different ecolodges, while some have narrow applicability. Only one of the ecolodges in the sample installed a desalination plant, which was a feasible solution in its context (location close to a seaside), but would not be a relevant initiative for a lodge in a different location. Similar observations were made concerning biodiversity conservation, where the practice of reimbursing local population for livestock killed by predators to protect these predators applies only to a particular context.

Sustainability: community development

Analysis of ecolodge communications addressing local community development led to the identification of five major themes and four sub-themes (Table 4). Most of the practices focused on three major themes: preserving socio-cultural heritage, improving social wellbeing and contributing to the local economy. Preservation of socio-cultural heritage was mainly done through architecture, but also local décor, arts, traditions and practices. Additionally, lodges protected heritage by getting involved in specific activities, outside the regular lodge operations, explicitly aimed at heritage preservation. These practices included raising awareness, organizing heritage chapters and supporting community groups focused on heritage preservation. Practices aimed at the second central theme in community development, improving social wellbeing, focused on education, health, quality of life and employee wellbeing. They reflected different degrees of lodge involvement. Some of the practices required less direct involvement or interaction with community members, such as providing financial support to sponsor teams or pay teacher salaries or providing in-kind donations, such as school supplies or medical equipment. Other initiatives meant that a lodge was playing a more active role in the community life, for example, by providing mobile health clinics, running education programs, or developing infrastructure together with locals.

Concerning the theme of contribution to the local economy, six major types of best practices emerged from data analysis. They ranged from hiring and purchasing practices focused on local people and goods, to more active involvement with local business community through representing and assisting local people in setting up businesses. The remaining two themes within community development focused on the inclusion of local people into lodge operations and

Table 3. Environment and conservation best practices adopted by the leading ecolodges.

Themes and sub-themes		Best practices
Sustainable design and construction		• Local materials: ensuring only local materials are used • Integration with nature: blending with the environment, site-sensitive development with minimal or no landscaping, tree cutting, or earth movements, small lodge size to reduce impacts • Environmentally-friendly or sustainable materials: renewable, natural, reclaimed, recycles, sustainably-sourced, or certified materials • Sustainable construction and energy efficient design: natural ventilation system, passive cooling design, open concept, window screens, materials that counteract hot summer temperatures and cooler winter temperatures, using hand tools during construction to minimize land disturbance • Location choice: taking into account effects on soil, water table, vegetation, animal movements, visual and aural implications
Environmentally-friendly lodge operations	Water management	• Waste water treatment, reuse and water purification: water purification system to supply clean drinking water, bio-filter systems for gray and waste water, construction of a wetland to treat the waste water from the kitchen and laundry before it enters the nature, gray water irrigation • Water conservation: low-flow shower heads, towel replacement policies, low water use toilets, low water consumption dishwashers and washing machines, leaks reduction, disposable and compostable plates, asking guests to bring own linens • Water harvesting and storage: rain or snow harvesting for laundry and irrigation, producing own water using desalination plant, using water from springs or boreholes
	Energy management	• Use of renewable energy to power the lodge: solar, wind, geothermal power, alternative fuels (e.g. olive pit charcoal) to light, heat or cool the lodge • Energy conservation: low-wattage light fittings, providing no electricity in the rooms, use of lighting sparingly, no air conditioning, energy efficient appliances, using only essential electrical appliances, air drying the laundry • Solar water heating: used for showers and pool
	Waste and supply chain management	• Waste reduction: minimize or eliminate single-use plastic, packaging, batteries; buying in bulk, minimizing food waste (e.g. adoption of a nose-to-tail philosophy), soap and shampoo served by dispensers, donating reusable linen bags to community, providing reusable water bottles or cups (stainless steel, ceramic, clay), complimentary filling stations • Sorting, reusing, recycling waste: recycled materials used to make slippers, decorations, souvenirs, planters; plastic and glass bottles used as building or crafts and arts materials; cans and boxes given to locals for reusing • Organic waste disposal: composting; feeding waste to local livestock • Sustainable supplies: using biodegradable and toxic free cleaning products, hand soaps, bathroom amenities, organic spa products; using supplier code of conduct • Sewage treatment: composting toilets, turning waste into biogas used as cooking fuel, using animal waste as fertilizer or building material • Buying sustainable food: sourcing organic, free range, fair trade and sustainably farmed produce, organic fruit and vegetables, seasonal food
	Pollution and carbon footprint reduction	• On-site food growing: herb and vegetable garden, organic garden, operating a greenhouse or a farm (e.g. shade-grown coffee farm, sustainable shrimp farm), keeping animals to provide eggs and milk • Reduce air pollution and carbon emissions: minimizing use of fossil fuels and carbon footprint, offering charging stations for electric vehicles for guests, reducing transportation by car-pooling by staff, using public transport, providing employee accommodation on-site • Reduce noise pollution: modification of boat engines, electrical vehicles for silent safaris, e-boat with solar panels, no motorized activities, not allowing helicopter landing on the property, not using generators, using solar powered pump for water supply to avoid disturbing the surrounding natural habitats • Reduce water pollution: no discharge of waste from boats into a lake, preventing sheep and cattle from neighbouring farms from gaining access to a lake • Reduce light pollution: turning lights off, using less lighting
Biodiversity conservation		• Sustainable land management and use of natural resources: designating land as conservation area, establishing marine protected area, responsible use of natural resources; reserve management plans • Restoration and preservation of wildlife: animal rescue, reintroduction of animals, such as mega-herbivores, outreach to prevent locals from killing predators, establishing wildlife sanctuaries

(*Continued*)

Table 3. Continued.

Themes and sub-themes	Best practices
	• Reforestation, re-vegetation, weed control: planting native/indigenous trees and plants, weeding non-native plant species, growing tree samplings in a nursery, planting drought resistant species • Collaboration with conservation organizations and public sector: membership in sustainable development and biodiversity conservation groups, opposing inappropriate development or land use, expanding protected public lands, community conservation campaigns (e.g. to remove power boats from a lake) • Financial contributions to conservation: percentage of income or profit, fixed amount per guest-stay, lump sum payments • Reducing impact from tourist activities: cap on tourism development, trails preparation with minimal impacts, limiting the number of visitors and excursion group size, prohibiting diving, banning the feeding of wildlife, vehicles prohibited from driving off-road • Law enforcement and responders support: anti-poaching operations removing snares, logistical and financial support to local anti-poaching activities, monitoring of poaching activity, purchase of a fire-fighting unit • Garbage collection in the natural areas: by staff, resident environmentalists, eco-tourist volunteers

minimizing negative social impacts, both of which were less prominent. When discussing the involvement of local people, two major practices emerged: hiring local people during lodge design, construction and operations, or ensuring local ownership and management of the lodge. There was little discussion of engaging locals in identifying essential sustainability issues, despite the importance of this aspect highlighted in the literature. Similarly, when discussing social impacts, lodges focused mostly on benefits and provided a limited discussion of potential negative impacts and their efforts to minimize these impacts. One notable example of a practice aimed at reducing negative social impacts was limiting the number of participants taking part in activities to minimize disturbing locals, implemented at Apani Dhani Ecolodge.

Education and research

The analysis of content devoted to education and research revealed four themes: learning opportunities for guests, influencing local attitudes, creating opportunities for visitors to get involved, and supporting research (Table 5). The best practices within the theme of providing learning opportunities for visitors were focused on the environment or local culture. In both cases, the practices included low-involvement activities, such as providing information in the form of books, displays or materials placed in guest rooms. At the same time, they also involved higher-involvement activities such as guided tours and direct interactions with local community members. The learning opportunities covered many diverse topics within science and nature, environmental stewardship and local culture and traditions. Guests could learn about local flora and fauna, biomes, interactions among different species, life cycles of animals and plants, as well as many aspects of environmental stewardship. Culture-focused education opportunities included topics such as history, traditional lifestyles, arts, customs, celebrations, medicine and beliefs.

In addition to educating guests, lodges in the sample placed considerable effort towards educating locals and influencing their attitudes to conservation. Ecolodges promoted the idea that ecotourism and nature tourism can provide better career and income-generating opportunities for locals compared to unsustainable resource exploitation through poaching, illegal mining, logging, or slash-and-burn farming. They further demonstrated to residents and landowners that wildlife and land conservation can enhance the attractiveness of the area to visitors and therefore lead to improved quality of life and additional benefits for the community. Best practices within this theme also included creating environmental education opportunities for locals, staff and guides. Topics included ecosystem protection, waste management, poaching, sustainable living, renewable energy and impacts of ecolodge operations. The remaining two themes focused on providing opportunities for guests to get involved in the community and in the sustainability-related efforts undertaken by the lodges as well as supporting research.

Moral imperative

Concerning moral imperative, five themes were identified (Table 6). The first theme, commitment to sustainability or ecotourism, involved best practices such as providing mission or values statements, codes of conduct and environmental management plans and policies. The second theme focused on ensuring third-party recognition and included sustainability or

Table 4. Local community development best practices adopted by the leading ecolodges.

Themes and sub-themes		Best practices
Preserving socio-cultural heritage		• Traditional or locally-inspired architecture, building techniques, materials: lodge based on refurbished or reassembled old traditional homes, ancient construction techniques (e.g. applying Buddhist technique without use of nails) • Incorporating traditional art and culture into the lodge: traditional and local art, artefacts, tools and equipment (e.g. canoes, candles) displayed in or used at the lodge, décor inspired by traditions of indigenous cultures (e.g. traditionally painted ceilings), hosting traditional events and performances by local musicians, dancers, performers of traditional art, implementing local customs such as drawing patterns with rice flour outside homes, menu announcements in local language • Native/traditional dishes: kitchen equipped with traditional ovens to prepare native dishes, traditional meals, indigenous herbs and vegetables incorporated into dishes • Educating and supporting locals with respect to heritage: heritage awareness workshops, support for arts organizations, hosting and organizing art camps and workshops for local artists, supporting students studying traditional music, coordinating activities of a heritage organization chapter, traditional musical instrument lending library for locals
Social wellbeing	Education	• Support local education: finance local school construction, renovations and maintenance (gym, library, kitchen, dining hall, administration buildings), provide school supplies and technology (e.g. e-readers), provide nutrition programs for students, pay teacher salaries, build houses for teachers, sponsor English classes and school sports teams • Support education beyond local school: scholarships for local students to attend culinary, music and performing arts programs, contributions to secondary and university school fees, programs to assists indigenous youth in obtaining education and finding stable careers, running education programs
	Health	• Health education: programs and camps aimed at girls, donations to local schools to create organic vegetable plots and greenhouses, cooking classes to promote healthy eating, health education in local schools, educating community on immunization and family planning • Health services: funding health personnel, buildings (e.g. maternity clinic, healthcare center), ambulance, surgeries for children in the community, medical supplies and equipment; providing mobile clinic, providing eye exams for community members, providing lodge vehicles to transport community members to hospital
	Quality of life	• Improved access to natural resources: supplying water from onsite well for animals belonging to nomadic herders, supplying water storage tanks and drinking water filtration systems to community; assisting locals with transportation and storage of water; supporting renewable energy projects; project extending gas piping to cooking huts in the village to make use of biogas for cooking and heat; donating fuel efficient stoves; providing food waste or grass for local livestock; allowing locals access to lodge grounds to collect firewood, walk to school or feed animals • Supporting social causes and local activities: donating food, clothes, linens, cutlery for orphanages, children centers, seniors' homes; hosting physically-challenged children at the lodge and organizing fun activities for them, providing library and Internet access for the local community, organizing and sponsoring sporting and cultural events, children's clubs, teams • Enhancing infrastructure and safety: running environmental NGO to collect garbage in the community, funding garbage disposal system, building and sponsoring a community recycling center, supporting law enforcement to fight crime
	Employee wellbeing	• Job permanency and salary: salaries above the national average or statutory requirements, full time and all-year employment as opposed to seasonal only, fair working conditions • Benefits: life insurance, access to company doctor for employees and their families including children and parents, staff trips • Employee training and professional development: opportunities for advancement, job training, training in business administration, accounting, marketing, sales, customer service, food and drink preparation, tourist guiding, English and other foreign languages, reading, writing, mathematical skills, first aid, rescue services, cartography, navigation, geography
Contribution to local economy		• Hire local people: local staff, guides, management; hire people from remote, rural communities; create opportunities for first-ever formal employment • Sourcing goods and services locally: furniture, arts, crafts, pottery, candles, soaps, activities offered to guests, contracting skilled and semi-skilled labour to local village to build decks, walls, benches; using local transportation companies for boat, air and ground transfer • Purchasing local food: support for fish farms and agriculture development, purchase of fish and marine products directly through local traders working with community fishers • Creating opportunities for local businesses: opportunities for selling local goods, crafts, food at the lodge shop or to lodge visitors, selling excursions and activities led by locals • Representing local businesses and supporting local business organizations: representing local business in discussion with local and regional government, industry groups • Support in setting up a business: financial and administrative support for small business development, for example family farms, horseback riding or mountain biking businesses; planning sustainable businesses for local producers

(Continued)

Table 4. Continued.

Themes and sub-themes	Best practices
	and sponsoring training and supplies needed for their operations, such as seeds, irrigation system, tools, bee hive boxes; providing no-interest loans to help start a business
Involving local people in planning, design, construction and operations	• Design/construction by the local/indigenous community: hiring local craftsmen, artisans, architects • Local or indigenous ownership or management: all administrative and operative staff native, operating as a partnership or joint venture between a tour operator and local community, employees are stakeholders of the business and get percentage of earnings
Minimizing negative social impacts	• Tour size management: limiting the number of participants in activities offered by local community to facilitate contact with locals and to disturb them as little as possible

environmental awards, certifications and external audits. The accountability theme included best practices in performance reporting. While the ecolodges provided a significant amount of information about the specific initiatives they implemented, there was much less information about the actual performance achieved and the outcomes of the initiatives.

In many cases, the measures reported focused on the scale of the initiatives, for example, by indicating the number of trees being planted, and not on the outcomes. In some cases, the best environmental and conservation practices led to 100% self-sufficiency concerning water or energy needs or complete elimination of plastic bottle and straws. More often, the performance information was limited or vague. The theme of sharing profits with the local community included situations when the community ran the lodge. Therefore all its revenues were reinvested into community projects or when the lodge committed to a fixed amount or percentage of revenues as a contribution to the community. Finally, best practices within the theme of contributions to social justice, focused mostly on gender equality through the facilitation of women's economic activities but also included supporting refugees or improving access to the market for farmers.

Discussion and conclusion

This study is the first to use a sample of leading ecolodges from around the world to identify ecotourism-related best practices. Although other studies have investigated ecolodges according to the use of defining characteristics of ecotourism (e.g. Lai & Shafer, 2005), these studies have been restricted by scale and focus. Furthermore, the present study avoided the inclusion of operators who self-declared themselves as an ecolodge, in favour of independent third-party designations. The purposeful selection of ecolodges from Mehta (2010), World Travel Awards (2007-2019) and the National Geographic's Top 50 Ecolodges (Christ & Siber, 2008) was instrumental in this selection.

The majority of ecolodges in the sample stated that they are mainly following ecolodge principles as established in the literature. Given that the sample consisted of leading ecolodges, this finding was expected and confirmed that the sample chosen for this study is a suitable one to study best practices. The data analysis provided several main insights into the nature of these best practices as implemented by the leading ecolodges. Several themes, sub-themes and best practices have been identified and provide a valuable benchmark for ecolodges aiming to improve their sustainability. The most developed themes show many different best practices and examples that ecolodges can learn from and implement to improve their operations. At the same time, the data analysis has shown that some themes are less represented with only a few specific practices. Notably, although many initiatives were aimed at reducing negative environmental impacts of ecotourism, there were hardly any initiatives aimed at reducing its negative social impacts. When discussing relationships with local communities, ecolodges focused mostly on explaining the social benefits from their operations. They avoided discussion of negative impacts that tourism activity could bring to the area, such as overcrowding, the commodification of culture or increased crime. Limited content related to reducing negative social impacts may indicate a much higher priority given to social benefits, but also points to increased risk of overlooking potential social problems that can emerge during ecolodge operations.

Similarly, ensuring accountability was addressed in a limited way, mainly by reporting performance of an ecolodge on its website. The performance reporting varied greatly in the sample, with some ecolodges providing detailed reports with several different measures and others providing minimal information on an exact time

Table 5. Education and research best practices adopted by leading ecolodges.

Themes and sub-themes		Best practices
Learning opportunities for tourists	Environment-focused	• Science/nature education: guided nature tours, interpretive displays, interpretation materials and books in guest rooms or a library, natural history centre, science presentations by naturalist guides, resident astronomer, resident marine biologist • Tours focused on sustainable practices at the lodge: tours of the back of the house, guided tours of an organic garden, greenhouse or farm; agro-tourism based tours to teach about reforestation and aquatic and organic farming practices at the lodge, tours about the design inspiration and local sourcing; visiting animal sanctuary • Educating visitors about environmental stewardship: providing books on environmental stewardship in lodge's library, providing environmental educational materials in guest rooms; on-site instruction or website section on minimizing impacts when travelling or staying at a lodge; asking guests to participate in sustainability efforts at the lodge; encouraging travellers to participate in non-polluting activities such as walking, cycling, or camel cart trips; explaining and proposing carbon offsetting to guests; educating guests about the lodge's code of ethics
	Local culture-focused	• Educational materials about local culture: providing books, magazines, presentations or documentaries about local culture, art, heritage or history; placing information about the region and its culture in guest rooms; providing advice for travellers explaining local culture and desired behaviours; awareness centre, exhibitions or displays of cultural collections • Tours and visits to cultural and heritage attractions: tours to archaeological, historical, or rock art sites; visits to villages, churches, monasteries, heritage museums, markets, tea shops, and fields; opportunity to watch local way of life, local artists performing cultural dances and songs, local festivals, celebrations, ceremonies or craftsmen working on their projects (making paper, beads, jewellery); guided tours explaining heritage, culture, traditions, sharing local legends, myths and beliefs, teaching about medicinal plants, Buddhist philosophy, heritage protection • Local community-led activities and interactions: workshops/demonstrations led by locals on traditional arts, crafts, house construction, food preparation, hunting techniques, honey collecting, beading, pottery, dyeing of clothes, basket weaving, medicine, archery, falconry, makeup, making leather, candles, tents, dairy products, or ropes; dream interpretation by the elders; visiting or spending a day with local/indigenous family or community members to experience their day to day life and activities (farming, fishing, working as a shepherd); participating in local events (ceremonies, bonfire, games, dances, rituals)
Influence local people and staff attitudes to conservation and ecotourism		• Demonstrating alternative economic activity and employment: promoting employment as guides, lodge staff, rangers, artisans, para-biologists, workshop leaders; demonstrating the value of biodiversity conservation (e.g. rhino conservation, predator protection) in generating income through tourism, encouraging local landowners to expand conservation area • Environmental education for local community: free environmental education programs, forest immersion activities, eco summer camps and eco-clubs, and rhino watch trips for school children; open days at the lodge for schools to learn about the importance of conservation; sustainability-oriented scholarship programs; free environmental workshops and events for teachers, fishers, farmers, other community groups and government officials • Environmental education for staff and guides: guides school teaching the philosophy of nature conservation and sustainability, training staff on the responsible use of resources, recycling, environmental best practices, wildlife rescue and supervision of visitors on the trails
Opportunities for tourists to get involved		• Volunteering opportunities: tree planting, wildlife monitoring, site rehabilitation, weeds and non-native species control, reforestation, coffee picking, teaching English and arts, gardening, beach cleaning, repairing fences, website and social media management, egg collecting, milking cows, separate opportunities for medical professionals and educators • Financial support: encouraging donations to lodge's foundation or the community, recommending a donation to a school after a school visit, encouraging guests to support projects by providing them with a detailed documentation on the organizations and contribution from the lodge • Bringing items for community: participating in Pack for Purpose or Pack-a-Pound initiatives, which encourage visitors to bring and donate supplies for educational programs or useful items for community in their luggage • Special activities/tours focused on conservation and community development: data collection, assisting conservation team with monitoring and tracking animals, de-horning rhinos, accompanying the anti-poaching team to survey and monitor the conservancy, reef monitoring, allowing guests to install and review camera traps, visiting local health clinics or schools, travelling with the mobile clinic to distribute medications and health advice • How you can help or get involved section on website: encouraging guests to visit, donate money or listed items, providing information about volunteer opportunities, providing links to donate
Supporting research		• Lodge-conducted research: forest, wildlife, reef monitoring; wildlife and bird counts; science laboratory with a resident wildlife project coordinator; wildlife research centre; use of photo-trapping (motion sensor cameras) to understand and record animal populations; use of a mini-drone to capture high-resolution imagery of the rainforest canopy; conducting socio-economic studies

(*Continued*)

Table 5. Continued.

Themes and sub-themes	Best practices
	• Logistics, management and in-kind support for external research projects: providing logistical support, equipment, human resources, fuel, vehicle servicing, access, property use for researchers and conservation groups; accommodation and food for scientists; hosting international experts • Sponsoring research projects: on biodiversity (e.g. birds, mammals, reptiles, fish, butterflies), water resources, soil contamination; financial support for construction of research lab station; supporting research through a foundation • Collaboration with conservation and academic organizations: partnerships, participation in conferences to share best practices, organizing conferences

frame. The same communications were directed at many different stakeholders, such as visitors, employees or business partners, pointing to the need for further development of best practices in this area. Other themes with a limited range of best practices were reducing noise and water pollution. Although highly relevant for many ecolodges, given their location close to water bodies and in ecologically sensitive areas, this aspect was addressed by very few specific initiatives, suggesting low perceived importance and materiality on the part of ecolodges. This further demonstrates the need for more comprehensive coverage of a full range of best practices to emphasize the three pillars of sustainability in ecolodges fully. Further research could investigate the underlying reasons and implications of limited efforts to addressed the above themes.

Additionally, the absence of best practices aimed at some of the issues identified in the conceptual model was noted, such as incorporating stakeholders' feedback in decision-making. We identified stakeholder theory (Freeman, 1984) as a core perspective in the relationship between an ecolodge and other stakeholders pulled into the sphere of influence of ecolodges. Key to this theory is the need to engage stakeholders cooperatively and collaboratively in addressing the needs of multiple actors. However, none of the ecolodges discussed best practices involving stakeholders in the identification of sustainability priorities or in incorporating stakeholder feedback. As a result, the current study helps identify areas where additional work is needed to develop a complete set of best practices and help improve sustainability of the ecolodge sector.

Table 6. Moral imperative best practices adopted by the leading ecolodges.

Themes	Best practices
Commitment to sustainability or ecotourism	• Stating commitment: providing mission statement, our philosophy statement, providing code of conduct, stating values • Providing formal policies: providing environmental policy, environmental management plan
Seeking external recognition for sustainability efforts	• Certification: undergoing process to obtain third-party environmental or ecotourism certification • Environmental or sustainability awards: gaining recognition through awards received from different ecotourism or conservation organizations • External audit: arranging for an external environmental audit carried out on an annual basis
Accountability	• Reporting performance: by providing performance measures on website, by publishing a newsletter including information on obtained results and outcomes, by issuing annual reports
Sharing profits with the community	• A fixed amount or percentage of revenues donated to the community: supplementary fees directly given to local families, 5% levy included in guests' accommodation bills funnelled back to the villagers, 50% of proceeds from 'day in a life of local family' tours going directly to visited families, a bed-night levy for conservation projects • Lodge 100% owned and operated by the community: all revenues going towards supporting the community and preserving their culture and protecting the environment, revenues reinvested into community projects such as renewable energy, education and health care
Contributions to social justice	• Supporting gender equality: training women to become camp guides or members of security team; offering employment for women in running cooking classes or creating beadwork; and supporting women-run social enterprises making pottery, candles, beach baskets, towels or providing laundry services • Commitment to equality: supporting immigrants and creating opportunities for them, elimination of discrimination, improving access to markets

An additional issue that emerged during the analysis was the changing importance and applicability of best practices. There were several instances when a best practice in one context was no longer a best practice after the context changed. One example is sourcing goods and services locally. For most ecolodges that adopted this practice, it was viewed as a positive action that supports community by creating demand for its products and generating revenue for community members that can then be spent on education, health, food or other local needs. However, one ecolodge in Kenya, Campi ya Kanzi, chose not to purchase locally and defended its decision as the more responsible one. The justification provided by the ecolodge highlighted the lack of sustainable farming practices in the area and the risk that local purchasing would contribute to unsustainable farming and further environmental damage. The lodge chose to bring in environmentally-responsible goods from a further distance despite higher logistical costs to avoid adverse effects of local purchasing, although at the same time it meant less support for the local community.

Another example was the practice of using local and natural materials in lodge construction. Viewed by the majority of the lodges as a good environmental practice, it was not entirely implemented by Canopy Tower ecolodge. Instead, an old military installation (radar tower) was transformed into an observation centre and a lodge. This application is an example of reuse, where an ecotourism use was applied to a surplus defence installation and was preferred to construct a new lodge. The above examples demonstrate the challenge with identifying best practices and promoting their widespread adoption in the sector.

Additionally, an important issue to be addressed is interactions between different best practices and the outcomes and performance achieved by the ecolodge. Pursuing some initiatives, for example, 100% community ownership or high-cost environmental investments may make it more challenging to achieve financial goals and sustainability for the operator. These observations confirm findings from previous studies that the application of best practices to another setting is not straightforward and can be problematic (Bretschneider et al., 2005). As proposed in the conceptual model, best practices and their effects, including financial performance, can be significantly affected by several external and internal factors relevant to each ecolodge setting and its particular human and ecological circumstances.

In closing, there are limitations to our study. Firstly, as explained in the methodology section, the purposive sampling method was chosen to ensure that the sample consists of industry leaders that exhibit superior performance and would be suitable properties to study best practices. However, this sampling method also introduced a bias towards high-end lodges, since National Geographic lodges are among the most expensive lodges in the world and lodges which are recipients of World Travel Awards needed to have sufficient funds to go through the assessment process. As a result, some smaller, lower-end properties could have been excluded, despite also being industry leaders. It is essential to recognize this bias when interpreting the results of this study. Next, the identification of best practices was based on the content analysis of communications published by the ecolodges on their websites and, as such, faced limitations related to self-reporting. Namely, the data analyzed for this study represents the perspectives of ecolodge owners or managers and may be affected by selective reporting. As a result, the actual on-the-ground policies and practices may differ from the ones stated on the website, and a third-party assessment of ecolodge's practices or decisions could result in different observations than the one presented by lodges themselves. Finally, we identified the best practices implemented by the ecolodges in our sample, but there is the possibility that additional practices that influence ecolodge sustainability performance were not identified in this study.

To overcome these limitations, future research on ecolodges could collect primary data through field-based or virtual interviews and surveys of owners, staff and customers in order to better understand the practices implemented and the effect of the internal and external factors on the choice of these practices and resulting performance, as proposed in our conceptual model. Primary data-based case studies including observation, interviews and surveys on motivating factors and barriers to implementation of best practices, would further create a deeper understanding of these sites. Additionally, longitudinal case studies could explore changes in best practices over time in response to changing environment and stakeholders' expectations and provide guidelines for ecolodges considering implementation of these practices. We argue that ecolodge operators may advance their philosophical thinking and operational behaviours through closer adherence to the U.N.'s Sustainable Development Goals in building a stronger connection to CSR. Although the investigation of these 17 global goals (U.N., 2020) was not the focus of our study, future research should endeavour to make this connection in areas such as no poverty, zero hunger, quality education, gender equality, good health and wellbeing, decent work, reduced inequalities, peace and justice, and partnerships. This connection is critical, given that ecolodges should be sustainable by nature should actively strive to raise the bar in redefining the essence of best practice.

Disclosure statement

No potential conflict of interest was reported by the author(s).

References

Aguinis, H., & Glavas, A. (2012). What we know and don't know about corporate social responsibility: A review and research agenda. *Journal of Management*, *38*(4), 932–968. https://doi.org/10.1177/0149206311436079

Blangy, S., & Mehta, H. (2006). Ecotourism and ecological restoration. *Journal for Nature Conservation*, *14*(3), 233–236. https://doi.org/10.1016/j.jnc.2006.05.009

Bretschneider, S., Marc-Aurele, F., & Wu, J. (2005). "Best practices" research: A methodological guide for the perplexed. *Journal of Public Administration Research & Theory*, *15*(2), 307–323. https://doi.org/10.1093/jopart/mui017

Carlsen, J., Getz, D., & Ali-Knight, J. (2001). The environmental attitudes and practices of family businesses in rural tourism and hospitality sectors. *Journal of Sustainable Tourism*, *9*(4), 281–297. https://doi.org/10.1080/09669580108667403

Carroll, A. B. (1979). A three-dimensional conceptual model of corporate social performance. *Academy of Management Review*, *4*(4), 497–505. https://doi.org/10.5465/amr.1979.4498296

Chan, J. K. L., & Baum, T. (2007). Motivation factors of ecotourists in ecolodge accommodation: The push and pull factors. *Asia Pacific Journal of Tourism Research*, *12*(4), 349–364. https://doi.org/10.1080/10941660701761027

Christ, C., & Siber, K. (2008). 50 top ecolodges: The most earth-friendly retreats in the world's most spectacular wilds. *National Geographic Adventure*, *10*(9), 104–108.

Coles, K., Fenclova, E., & Dinan, C. (2013). Tourism and corporate social responsibility: A critical review and research agenda. *Tourism Management Perspectives*, *6*, 122–141. https://doi.org/10.1016/j.tmp.2013.02.001

Creswell, J. W., & Plano Clark, V. L. (2007). *Designing and conducting mixed methods research*. SAGE Publications.

Deegan, C. (2002). Introduction: The legitimizing effect of social and environmental disclosures – A theoretical foundation. *Accounting, Auditing & Accountability Journal*, *15*(3), 282–311. https://doi.org/10.1108/09513570210435852

Donohoe, H. M., & Needham, R. D. (2008). Internet-based ecotourism marketing: Evaluating Canadian sensitivity to ecotourism tenets. *Journal of Ecotourism*, *7*(1), 15–43. https://doi.org/10.2167/joe185.0

Drumm, A., Moore, A., Soles, A., Patterson, C., & Terborgh, J. E. (2004). *Ecotourism development. A manual for conservation planners and managers. Volume II: The Business of ecotourism development and management*. The Nature Conservancy.

Epler Wood, M. (2002). *Ecotourism: Principles, practices and policies for sustainability, United Nations environment programme*. United Nations Publications.

Fennell, D. A.. (2020). *Ecotourism*. Routledge.

Fennell, D. A. (2001). A content analysis of ecotourism definitions. *Current Issues in Tourism*, *4*(5), 403–421. https://doi.org/10.1080/13683500108667896

Freeman, R. (1984). *Strategic management: A stakeholder approach*. Pitman.

Godfrey, P. C., Merrill, C. B., & Hansen, J. M. (2009). The relationship between corporate social responsibility and shareholder value: An empirical test of the risk management hypothesis. *Strategic Management Journal*, *30*(4), 425–445. https://doi.org/10.1002/smj.750

Gray, R., Kouhy, R., & Lavers, S. (1995). Corporate social and environmental reporting: A review of the literature and a longitudinal study of U.K. disclosure. *Accounting, Auditing & Accountability Journal*, *8*(2), 47–77. https://doi.org/10.1108/09513579510146996

Hakim, L., & Nakagoshi, N. (2014). Ecotourism and climates changes: The ecolodge contribution in global warming mitigation. *The Journal of Tropical Life Science*, *4*(1), 26–32. https://doi.org/10.11594/jtls.04.01.05

Hawkins, D. E., Epler Wood, M., & Bittman, S. (1995). *The ecolodge sourcebook for planners and developers*. The International Ecotourism Society.

Hendry, J. (2001). Economic contracts versus social relationships as a foundation for normative stakeholder theory. *Business Ethics: A European Review*, *10*(3), 223–232. https://doi.org/10.1111/1467-8608.00236

Herzberg, F., Mausner, B., & Snyderman, B. (1959). *The motivation to work*. Wiley.

International Finance Corporation. (2004). *Ecolodges: Exploring opportunities for sustainable business*. World Bank Group.

Jones, T., Harrison, J., & Felps, W. (2018). How applying instrumental stakeholder theory can provide sustainable competitive advantage. *Academy of Management Review*, *43*(3), 371–391. https://doi.org/10.5465/amr.2016.0111

Kim, S. Y., & Park, H. (2011). Corporate social responsibility as an organizational attractiveness for prospective public relations practitioners. *Journal of Business Ethics*, *103*(4), 639–653. https://doi.org/10.1007/s10551-011-0886-x

Kwan, P., Eagles, P., & Gebhardt, A. (2008). A comparison of ecolodge patrons' characteristics and motivations based on price levels: A case study of Belize. *Journal of Sustainable Tourism*, *16*(6), 698–718. https://doi.org/10.1080/09669580802397129

Kwan, P., Eagles, P., & Gebhardt, A. (2010). Ecolodge patrons' characteristics and motivations: A study of Belize. *Journal of Ecotourism, 9*(1), 1–20. https://doi.org/10.1080/14724040802140576

Lai, P. H., & Shafer, S. (2005). Marketing ecotourism through the Internet: An evaluation of selected ecolodges in Latin America and the Caribbean. *Journal of Ecotourism, 4*(3), 143–160. https://doi.org/10.1080/JET.v4.i3.pg143

Laugen, B., Acur, N., Boer, H., & Frick, J. (2005). Best manufacturing practices: What do the best- performing companies do? *International Journal of Operations & Production Management, 25*(2), 131–150. https://doi.org/10.1108/01443570510577001

Lii, Y. S., & Lee, M. (2012). Doing right leads to doing well: When the type of CSR and reputation interact to affect consumer evaluations of the firm. *Journal of Business Ethics, 105*(1), 69–81. https://doi.org/10.1007/s10551-011-0948-0

Lu, W., & Stepchenkova, S. (2015). User-generated content as a research mode in tourism and hospitality applications: Topics, methods, and software. *Journal of Hospitality Marketing & Management. 24*(2), 119–154. https://doi.org/10.1080/19368623.2014.907758

Mafi, M., Pratt, S., & Trupp, A. (2019). Determining ecotourism satisfaction attributes – a case study of an ecolodge in Fiji. *Journal of Ecotourism, 19*(4), 304–326. https://doi.org/10.1080/14724049.2019.1698585

Massi, M., & De Nisco, A. (2018). The Internet-based marketing of ecotourism: Are ecotourists really getting what they want? In M. A. Camilleri (Ed.), *Tourism planning and destination marketing* (pp. 161–182). Emerald Publishing Limited.

Mehta, H. (2007). Towards an internationally recognized eco lodge certification. In R. Black, & A. Crabtree (Eds.), *Ecotourism series, number 5: Quality assurance and certification in ecotourism* (pp. 415–434). CABI Publishing.

Mehta, H. (2010). *Authentic ecolodges.* Harper Collins Publishers.

Mehta, H., Baez, A., & O'Loughlin, P. (2002). *International ecolodge guidelines.* TIES.

Millar, M., & Sammons, G. (2006). A content analysis of Costa Rican Ecolodge websites. hospitality management. Paper 11. http://repository.usfca.edu/hosp/11.

National Geographic. (2019). Unique lodges of the world: Sustainable tourism in action. Sustainable tourism impact report. Retrieved from: https://issuu.com/nationalgeographicexpeditions/docs/lodges_report_6.21

Nelson, V. (2010). Promoting energy strategies on eco-certified accommodation websites. *Journal of Ecotourism, 9*(3), 187–200. https://doi.org/10.1080/14724040903144352

Nowaczek, A., & Smale, B. J. A. (2010). Exploring the predisposition of travellers to qualify as ecotourists: The ecotourist predisposition scale. *Journal of Ecotourism, 9*(1), 45–61. https://doi.org/10.1080/1472404090288352

O'Rourke, A. (2003). A new politics of engagement: Shareholder activism for corporate social responsibility. *Business Strategy and the Environment, 12*(4), 227–239. https://doi.org/10.1002/bse.364

Osland, G. E., & Mackoy, R. (2004). Ecolodge performance goals and evaluations. *Journal of Ecotourism, 3*(2), 109–126. https://doi.org/10.1080/14724040408668153

Plaza-Úbeda, J. A., de Burgos-Jiménez, J., & Carmona-Moreno, E. (2010). Measuring stakeholder integration: Knowledge, interaction and adaptational behavior dimensions. *Journal of Business Ethics, 93*(3), 419–442. https://doi.org/10.1007/s10551-009-0231-9

Sanders, E., & Halpenny, E. (2001). *The business of ecolodges: A survey of ecolodge economics and finance.* The International Ecotourism Society.

Sangpikul, A. (2010). Marketing ecotourism through the Internet: A case of ecotourism business in Thailand. *International Journal of Hospitality & Tourism Administration, 11*(2), 107–137. https://doi.org/10.1080/15256481003732782

Scheyvens, R. (1999). Ecotourism and the empowerment of local communities. *Tourism Management, 20*, 245–249. https://doi.org/10.1016/S0261-5177(98)00069-7

Sen, S., & Bhattacharya, C. B. (2001). Does doing good always lead to doing better? Consumer reactions to corporate social responsibility. *Journal of Marketing Research, 38*(2), 225–243. https://doi.org/10.1509/jmkr.38.2.225.18838

Stemler, S. (2000). An overview of content analysis. *Practical Assessment, Research, and Evaluation*, 7(17), 1–6. https://scholarworks.umass.edu/pare/vol7/iss1/17

Stronza, A., & Gordillo, J. (2008). Community views of ecotourism. *Annals of Tourism Research, 35*(2), 448–468. https://doi.org/10.1016/j.annals.2008.01.002

The International Ecotourism Society (TIES). (1993). *Ecotourism guidelines for nature tour operators.* The International Ecotourism Society.

Tzschentke, N., Kirk, D., & Lynch, P. A. (2008). Going green: An examination of decisional factors in small hospitality operations. *International Journal of Hospitality Management, 27*(1), 126–133. https://doi.org/10.1016/j.ijhm.2007.07.010

United Nations. (2020). The 17 goals. https://sdgs.un.org/goals.

Wallace, G. N., & Pierce, S. M. (1996). An evaluation of ecotourism in Amazonas, Brazil. *Annals of Tourism Research, 23*(4), 843–873. https://doi.org/10.1016/0160-7383(96)00009-6

Weaver, D. B., & Lawton, L. J. (2002). Overnight ecotourist market segmentation in the Gold Coast hinterland of Australia. *Journal of Travel Research, 40*(3), 270–280. https://doi.org/10.1177/004728750204000305

Wight, P. (1997). Ecotourism accommodation spectrum: Does supply match the demand? *Tourism Management, 18*(4), 209–220. https://doi.org/10.1016/S0261-5177(97)00008-3

Wood, M. E. (2002). *Ecotourism: Principles, practices & policies for sustainability.* UNEP.

An ecofeminist approach to ecotourism development

Margaret Byrne Swain and Melissa Tandy Byrne Swain

Abstract

This paper raises the question of what an ecofeminist approach may contribute to ecotourism development. It addresses shared discourse in the international literatures of these distinct ecologically based fields of study. Concepts such as sustainability and environmental justice inform both bodies of work. We illustrate this through the words of ecofeminist Vandana Shiva and ecotourism researcher Tej Vir Singh. Uses of ecofeminist-based gender analysis in ecotourism development projects are examined, drawing from the observations of various researchers, and case studies from around the globe. In conclusion, it is argued that an ecofeminist approach asks critical questions for ecotourism development projects, highlighting issues of agency, gender awareness, and equity necessary for ecologically sustainable communities.

Introduction

A shared discourse of sustainability and environmental justice informs the international academic literatures on both ecofeminism and ecotourism. This article briefly addresses the parallel histories of these ecologically based interests and then explores how ecofeminism might engage ecotourism to fulfil some of their mutual rhetorical promises. Ecofeminism as a social movement differs greatly in organization and practice from ecotourism as a strategy for economic development. However, gender equity planning promoted by ecofeminism could be applied in ecotourism projects and thus operationalize ecological goals of both, as seen in case studies from around the globe. In conclusion, this paper argues that an ecofeminist approach raises critical questions for ecotourism development projects, highlighting issues of agency, gender awareness, and equity necessary for ecologically sustainable communities.

If we can agree that tourism is built of human relations and thus impacts and is impacted by global and local gender relations, then it makes sense to apply feminist analysis in tourism studies (Swain 2002). A catch for some researchers is the political nature of feminist movements engaging theory with practice to promote gender equity. For example, it may not seem reasonable to interpret the connections between tourism and gender in the way political scientist Cynthia Enloe did back in 1989. Enloe (1989: 41) flatly stated that '[m]en's capacity to control women's sense of their security and self-worth has been central to the evolution of tourism politics.' However, for Enloe and other feminists of varied political persuasions, the actions of tourism industry workers and the self-employed, from prostitutes to artisans, to organizers of alternative tourism for women, '...should be seen as political, internationally political.' Enloe predicted that '[m]ovements that upset any of the patterns of today's tourist industry are likely to upset one of the principal pillars of contemporary world power.'

One of the principle pillars of power today is the 'axis of eco', concern and contestation over the use, abuse, and conservation of natural environments. With the addition of an eco-prefix, a type of feminism is named which focuses on environmental concerns, and a type of tourism practice becomes associated with the ideology of environmental movements. Ecotourism and ecofeminism share a similar vocabulary, and are linked by gender matters in the analysis and practice of ecotourism. Ecofeminism needs compatible forms of community change to thrive, while ecotourism needs compatible strategies to promote sustainable communities. It appears to be a win-win combination.

Terms and Definitions

Both ecofeminism and ecotourism have a range of definitions, depending on a scholar's or practitioner's particular focus. However, a basic definition of ecotourism offered by Hall and Kinnaird (1994: 111) uses the term to '...cover tourism development which is sympathetic to, complements and/ or is employed as a vehicle for, conserving and sustaining natural and cultural environments and their resources.' Ecofeminism, of course, focuses on human interactions with the environment, as can be seen in the following, often-cited definition offered by Sturgeon (1997: 23): '[E]cofeminism is a movement that makes connections between environmentalisms and feminisms; more precisely, it articulates the theory that the ideologies that authorize injustices based on gender, race, and class are related to the ideologies that sanction the exploitation and degradation of the environment.' Three major issues are woven through ecofeminist literature: connections between the domination of women and the domination of nature, women's relationships with nature, and the role of women in solving ecological problems (Momsen 2000).

Like ecofeminisms, some concepts of ecotourism are mindful of inequalities basic to global systems that make sustainability unreachable for the poor in need of short-term survival strategies (Cater 1994: 85). Honey (1999) has recently pushed the agenda of ecotourism further by arguing for a definition that includes the promotion of human rights and democratic values. However ecofeminism and ecotourism are defined, they indicate change and responsibility. As Wall has noted (1996:2), the introduction of tourists to previously remote areas '...will place new demands upon the environment associated with new actors, new activities, and new facilities. This will involve the forging of new relationships between people and environment, between peoples with different life styles, and between a wide variety of forces for both change and stability.'

A key word in common, ecology, is a good place to start an analysis of these international academic literatures based on ecological interests. The Concise Oxford Dictionary (9th edition) defines 'ecology' as follows: 1. The branch of biology dealing with the relations of organisms to one another and to their physical surroundings; 2. (in full, human ecology) the study of the interaction of people with their environment. Both meanings are evident in ecofeminist and other environmentalist writings. Given the human basis of ecotourism, the second meaning underlies our discussion here. Other terms that occur in common include: sustainability, environmental justice, indigenous knowledge, local communities, protection, empowerment, human rights, and democracy.

History, Discourse, and Critique

We will briefly address the parallel histories and rhetorical promises of ecofeminism and ecotourism by holding an 'interview' in the words of two leading scholars. We should note that some of the earliest work still influencing debates came out of India. In terms of ecofeminism, there was a '...combination of developed feminist and environmentalist movements [that meant] a high profile for women's resistance to a number of environmental problems in India' (Sturgeon 1997:124). Environmentalism worldwide has historically interacted with other kinds of movements and continues to influence various alternative development concepts. Ecofeminism and ecotourism were named eco-topics of study in academic articles by the early 1990s. They claim a common heritage from the Brundtland Report of 1987, and the Rio Conference of 1992 that raised global consciousness of ecological concerns among researchers, governments, industries, and citizenry.

Tej Vir Singh (1992: 122–126; 1999), founder of the Centre for Tourism Research and Development, draws from his own work starting in the 1970s and that of many other scholars, to understand and promote sustainable tourism development. From his point of view, the basics of this development include conserving natural resources, deepening the visitor experience, and enhancing the social and economic well-being of the community. He relates these observations to his beloved Kulu Valley in the Himalayas. Here he notes that voluntary efforts, informed policy, and active participation of resident communities hold promise for future green tourism despite tremendous pressures on the landscape.

Vandana Shiva (1990, 1996, 1998), a physicist turned ecofeminist scholar and activist, draws from her early work about conditions in the Himalayan foothills as sources of inspiration and concern. She founded the Research Foundation for Science, Technology and Ecology in 1982. Her observations of the grassroots Chipko movement, women protesting and stopping deforestation by literally hugging trees in the Garhwal hills in India, have been a great resource for many other ecofeminists. Her current work focuses on globalization and its impacts specifically on women's lives and the natural environment, such as biopiracy. She has yet to write on tourism, but it would be an ideal topic for her.

A Dialogue between Ecotourism and Ecofeminism

In order to discuss both ecotourism and ecofeminism, we thought to create a dialogue in the form of an interview between Tej Vir Singh and Vandana Shiva using quotes from their work.

INTERVIEWER: *Prof. Tej Vir Singh, in terms of economic and environmental sustainability, why do you believe that it is so important to embark upon drastic modes of change within tourism as it is today?*

TEJ VIR SINGH: *"Whereas capital in other sectors of the economy can be lost and gained again but not in tourism. Once lost, this is lost for good. Ruined paradises are seldom regained." (1992: 118) Furthermore, "When tourism culture abruptly invades such sensitive environments, the negative impacts are bound to outweigh the positive ones. The degree of change will depend largely upon the type of tourism, the scale and pace of development, the philosophy behind it - and the resilience of the host community." (1999)*

INTERVIEWER: *Again, Prof. Singh, what parameters would you say are necessary for sustainable ecotourism?*

TVS: *"Scholars of tourism are enjoined to explore the possibilities of finding working model(s) for sustainable style or form of tourism that should be economically productive, socially responsible, and environmentally sound, involving participation of indigenous population. It should be a development that conserves and not consumes the environment, establishing tourism and environment compatibility. (1992: 120)*

INTERVIEWER: *Prof. Vandana Shiva, in terms of ecological development, what would be your main concerns?*

VANDANA SHIVA: *"The two big strands of my engagement and my thinking are on the economic front, the whole issue of globalization, and on the ecological front. A group of issues that are covered by biodiversity, the different species on this planet, that include human beings." (1996) "The paradox and crisis of development arise from the mistaken identification of culturally perceived poverty as real material poverty and the mistaken identification of the growth of commodity production as solving basic needs. But the growing ecological crisis and the ecological roots of poverty and the threat to survival indicate that this 'development' is leading to a shrinking pie for the peoples of the Third World. " (1990: 199)*

INTERVIEWER: *Prof. Shiva, insofar as tourism development can be deemed part of globalization, where do you see ecofeminism playing a key role?*

VS: *"Ecofeminism has emerged as a serious challenge to Cartesian reductionism and the ... 'masculine mode of knowing'. Globalization is, however, threatening to the ecological gains of the past few decades. It is, therefore, the defining context of our new engagements." (1998) "Development is very much like globalization, it's a word that is both constructed and has some kind of a natural connotation... Development comes from that domain of biological evolution. It has been applied to an economic domain of engineered, imposed paradigms of what's society's priorities should be. And exactly like globalization is made to look natural, development is made to look natural..." (1996)*

INTERVIEWER: *Prof. Singh, Prof. Shiva mentioned the "ecological gains" that have been made in the recent past. Whereas she intimates that ecology in general can both grow and diminish depending upon economic and social conditions, you have stated that it is very difficult to regain areas that have been diminished by tourism. How do you see ecotourism playing a role in either protecting or rebuilding touristic environments?*

TVS: *"[Mountain people in the past had 'habitat wisdom', which I love to call the sense of 'folk ecology' for their niche. They used their environment with parsimony; their livelihood pursuits were by and large ecological in nature, such as agriculture, forestry, pastoralism, or animal husbandry which kept the landscape sustainable." (1992: 115) "Practitioners of ecotourism are ordained to maintain the beauty and splendor of these environments, in all the gregariousness of bio-cultural diversities. These spectacular Earth-features should be carefully conserved to be passed on to posterity as the heritage of mankind, not turned into commodities to be consumed by tourism sharks. " (1999)*

INTERVIEWER: *Prof. Shiva, where do you see ecofeminism playing a role in the future of either protecting or rebuilding environmental areas?*

VS: *"I think, because men and women are victims of globalization in terms of disenfranchisement, joblessness, insecurity, the issue for me and for the feminist movement right now is for women ...to provide a feminist leadership for economic and social transformation. What that means really is leadership for men and women." (1996) "[P]eople have a right to live, they have a right to work, a right to know, a right to food, a right to shelter, they have a right to peace, they have a right to security. They have a right to imagine future generations will be living with some kind of predictable security: economic, social and cultural. (1996)*

INTERVIEWER: *Thank you both for your perspectives.*

Discussion

How then might ecofeminism engage ecotourism to fulfil some of the rhetorical promises of both interests? At one level, it would seem that there is little possibility for engagement. A strain of nostalgia for environmental authenticity evident in both literatures grows from essentialist thinking about what can and should be. However, essentialism can be turned upside down, made to open up connections rather than divide, as we shall see.

Ecotourism scholars have rarely evoked feminism as a possible way of theorizing and implementing change for conservation and equitable use of resources. Similarly, feminists have rarely raised the possibilities of 'good' ecotourism. Maria Mies, co-author with Shiva of the influential and much debated book *Ecofeminism* (1993:132), for example, expounds on links between the exploitation of the environment and the exploitation of human bodies. She names nature tourism and integrated rural tourism as sites of this degradation. Exceptions to this divide have been a few feminist tourism researchers working in rural, nature-based or agricultural tourism (Swain 2002).

The non-profit organization EQUATIONS has published a series of papers on its website about equitable tourism questions in India, including an article by Nina Rao (1997) on 'women and tourism.' Rao picks up on the theme of tourism and the environment, focusing on the possibility of rural women being displaced from their previous occupations with the replacement of agriculture by tourism work. She cautions that these conditions may result in women's loss of income from their expertise, and a lack of reemployment.

Regina Scheyvens (2000), in an overview of ecotourism ventures in the Third World, makes the case that although projects often profess to be 'gender neutral', they run the risk of disadvantaging and marginalizing local women. Changes in gender roles, relations, and access to resources may or may not empower women. Even if projects are designed specifically to assist women, the results may not go as planned if communities are unbalanced.

Barbara Dilly (2003:58–59), in her work on gender and ecotourism development in Guyana, draws similar conclusions from her research, without reference to either ecofeminism or gender in tourism literatures. She identifies several concerns regarding women in ecotourism projects, starting with the questionable sustainability of ecotourism projects that directly benefit women but do not integrate men. Dilly also calls for the evaluation of projects in terms of local gender relations, potential economic gain, and empowerment of both women and men, and an awareness that projects are gendered, which women and men experience differently. In her case study, Dilly (2003: 71) finds that even if the local population has a history of equitable gender relations, within the context of global capitalism it is difficult to develop non- exploitative infrastructures and promote just opportunities for women and men.

So, why has there been an overriding lack of gender perspective in environmental development projects, including ecotourism, despite the parallel literature and movement of ecofeminism? The work of Bina Agarwal (1997: 55–57) has raised this question and her work suggests that perhaps one problem is the re-entrenchment of gender inequalities under the guise of reviving 'authentic traditional ways.' She asks '...why has ecofeminism failed to provide a corrective? To what extent can it serve?' Ecofeminism as an ideology often romanticizes the past and obscures political economy factors underlying women's subordination and nature's degradation. Environmentalist theory and social movements have long agreed that effective management of local natural resources requires involvement of user communities. In tourism studies, Cohen (1992: 276–7) for instance has called for appropriate policies to shape interactions between mass and alternative tourism industry with local community participation. A remaining question is what community institutions should be used? Institutions, old or new, built on inequalities will of course perpetuate unequal participation and benefits, based on ideas of race, caste, ethnicity, class, or gender.

Another important question is one of agency. Agarwal (1997:72–3) asks if having an interest in changing something is enough to initiate the process of change. Citing examples in contemporary India, she finds little evidence that women's participation in environmental movements has furthered women's interests or changed gender relations. Continuing constraints on women's participation and effectiveness include (1997:77–78) women's double work burdens, official male bias, social constraints, and women's lack of recognized authority. For Agarwal, the critical factors are gender divisions of labour, power, and authority between women and men that need to be transformed, just as people transform notions of nature and human interaction with nature. The problem then is one of process (1997: 83), of figuring out how engagement in ecologically sound work could also serve the goals of equity. In terms of gender relations, Agarwal calls for strengthening women's bargaining power to increase their autonomy in the family and community. She (1997:85–6) names three interlinked aspects determining this power: (1) women's independent economic and legal status; (2) external support (NGOs, women's organizations, kin, friends, the State); (3) gender ideology.

Ecofeminism in Ecotourism

Gender equity planning in specific ecotourism projects may operationalize ecological goals of social movement and development. This paper illustrates this claim with brief observations of ecotourism development projects around the globe, looking at a range of sources for external support engaged with local communities. These sources of support are organized at transnational, state, and local levels. Transnationally, a great deal of activity can be currently found in Yunnan, China, including WTO planning and The Nature Conservancy's massive ecotourism development project. Here international teams of planners take into consideration natural and cultural environments, working with local communities and regional specialists (Eckholm 2001). Gender dynamics and the roles of women and men in creating sustainable tourism resources is an important, named concern. The province cooperates with international NGOs in this case.

The State may also cooperate with grassroots groups. As of 2002 in the US, a University of California Small Farm Center project has been proposed that would involve an activist organization, California Women for Agriculture, in a programme to promote ecotourism as an economic survival strategy in rural communities. Women's and men's roles in family farms and new possibilities in farm hospitality and local ecotourism are primary concerns (Jolly 2002). Ecotourism projects with a gender agenda are also being organized from the local level, with links out to State and transnational organizations. In Quintana Roo, Mexico, a Mayan village with help from a local philanthropist with international links, has been developing a self-help ecotourism project, showcasing their heritage while providing income. Community input has been along gender lines with distinct women's and men's roles and equitable, cooperative projects (Swain 2002).

Co-religious organizations are another base of ecotourism development, illustrated by the US/Romanian NGO Project Harvest Hope and 'pilgrimage tours' for US Unitarians to discover their religious roots in Transylvania (Keyes 2002). Ecotourism components include home stays in rural villages, participation in local work, and sharing of information on local environmental conditions. Equity is a primary concern of the NGO, which monitors the workload of community members in the project, supporting old gender divisions of labour as well as promoting new roles. Much of this project is similar to the Kanda Community case of voluntary tourism in the Himalayas described by Tej Vir Singh and Shalini Singh (2002). In this case as well a local leader took on the project of using tourism to promote equitable economic well-being and conserve the environment of his community. McGehee (2002) has nicely documented how the participation in alternative, volunteer-oriented tourism influences tourists to subsequently participate in related social movements. Tourism becomes a soft path into activism.

Conclusions

An ecofeminist approach raises critical questions for ecotourism development projects, highlighting issues of essentialism, agency, gender awareness, and equity necessary for ecologically sustainable communities. Noel Sturgeon has turned ecofeminist essentialism on its head, asking how belief in the intrinsic importance of the environment and women's place in it may be used to promote ecologically sound protection and equity in the use and management of natural resources. Rather than rejecting essentialism, Sturgeon (1997: 8-9) asks if 'feminization of nature contains resources for empowering women in a sexist society?' She (1997:149) points to Shiva's work identifying 'global environmentalism,' a discourse that blames the victims for environmental degradation, as a project of 'patriarchal capitalist maldevelopment.' In this discursive move, '...the oppositional accomplishment is to point out the sexism of the appeal to a generic mankind, and humanity, and to uncover, instead of universality, the connections between sexism, racism, imperialism, classism, and ...environmental exploitation.'

Sturgeon sees essentializing as discourse based on unequal power relations, which is the problem, not essentialist rhetoric (as metaphor, dramatization) *per se*. While it can maintain positions of privilege it can also produce 'oppositional consciousness.' Essentialist rhetorics in political collectives can be liberatory.

A recent work by Barbara Humberstone (2004) that does discuss ecotourism in conjunction with ecofeminism evokes a 'foundational re-conception of the woman/nature, man/culture binaries and non-human exploitation so prevalent in early Western thought' (2004: 125). Humberstone relates ecotourism to a particular view of ecofeminism, one that implies a 'common oppression of women and nature and the concern to transform this double subjugation' (2004:125). However, unlike Sturgeon's radical approach to essentialist discourse, Humberstone seems to buy into it. Despite the author's insistence upon considering issues of gender and ethnicity together with the development of ecotourism (2004:127), her conclusions do not appear to distance her from the limiting view that women are naturally more connected than men to the environment. Humberstone does not appear to engage the combined issues of ecotourism and ecofeminism in such a way that lends itself to praxis.

Once the inequalities are identified, as Agarwal spells out, then changes that promote greater equity in the life conditions of those doing environmental work will reinforce real connections between social and economic justice. As we saw in the words of Tej Vir Singh and Vandana Shiva, the nostalgia for environmental authenticity is a message to inspire work for the greater good, engaging ecologically sound work that serves the goals of equity.

References

AGARWAL, B. (1997). Environmental Management, Equity and Ecofeminism: Debating India's Experience. *The Journal of Peasant Studies* 25(4): 55–95.

CATER, E. (1994). Ecotourism in the Third World – Problems and Prospects for Sustainability. In Cater, E. and Lowman, G. (eds.) *Ecotourism: A Sustainable Option?* Chichester. John Wiley & Sons Ltd: 69–86.

COHEN, E. (1992). Alternative Tourism – A Critique. In Singh, T. V., Smith, V.L., Fish, M. and Richter, L.K. (eds.) *Tourism Environment: Nature, Culture, Economy.* New Delhi. Inter-India Publication: 269–277.

DILLY, B. (2003). Gender, Culture, and Ecotourism: Development Policies and Practices in the Guyanese Rainforest. *Women's Studies Quarterly* (3 & 4): 58–75.

ECKHOLM, E. (2001). A Holy Place in China Fights for its Life, Body and Soul. *The New York Times,* 10 June.

ENLOE, C. (1989). *Making Feminist Sense of International Politics: Bananas, Beaches and Bases.* Berkeley. University of California Press.

HALL, D. and KINNAIRD, V. (1994). Ecotourism in Eastern Europe. In Cater, E. and Lowman, G. (eds.) *Ecotourism: A Sustainable Option?* Chichester. John Wiley & Sons Ltd: 111–136.

HONEY, M. (1999). *Ecotourism and Sustainable Development: Who Owns Paradise?* Washington, DC. Island Press.

HUMBERSTONE, B. (2004). Standpoint Research: Multiple Versions of Reality in Tourism Theorizing and Research. In J. Phillimore and L. Goodson (eds.) *Qualitative Research in Tourism: Ontologies, Epistemologies and Methodologies.* London. Routledge: 119–136.

JOLLY, D. (2002). Personal communication.

KEYES, D. (2002). Personal communication.

McGEHEE, N. (2002). Alternative Tourism and Social Movements. *Annals of Tourism Research* 29(1): 124–143.

MIES, M. (1993). White Man's Dilemma: His Search for What He Has Destroyed. In Mies, M. and Shiva, V. (eds.) *Ecofeminism.* London. Zed Press.

MOMSEN, J. (2000). Gender Differences in Environmental Concern and Perception. *Journal of Geography* 99: 47-56.

RAO, N. (1997). Women and Tourism. http://www.equitabletourism.org/womenandchild.htm. Accessed on 24 July 2004.

SCHEYVENS, R. (2000). Promoting Women's Empowerment Through Involvement in Ecotourism: Experiences from the Third World. *Journal of Sustainable Tourism* 8(3): 232–249.

SHIVA, V. (1990). Development as a New Project of Western Patriarchy. In Diamond, I. and Orenstein, G. (eds.) *Reweaving the World: the Emergence of Ecofeminism.* San Francisco. Sierra Club Books: 189–200.

SHIVA, V. (1996). Interview. In V. Britto, *Talking About Development, an Interview Project.* http://www-personal.umich.edu/~fiatlux/td/vandana/vs-eng.html Accessed on 2 December 2002.

SHIVA, V. (1998). Message from the Founder. The Foundation for Science, Technology, and Ecology. http://www.vshiva.net/info.htm. Accessed on 2 December 2002.

SINGH, T.V. (1992). The Malaise of Mountain Tourism: Searching Strategies for Sustainable Development. In Singh, T.V., Smith, V.L., Fish, M. and Richter, L.K. (eds.) *Tourism Environment: Nature, Culture, Economy.* New Delhi. Inter-India Publications: 115–127.

SINGH, T. V. (1999). Keep the Sharks out of the Mountains. *Our Planet* 10(1). Nairobi, UNDP. http://www.ourplanet.com/txtveresn/101/ singh.html. Accessed 2 February 2002.

SINGH, T. V. and SINGH, S. (2001). The Emergence of Voluntary Tourism in the Himalayas: A Case of Kanda Community. Paper presented at the IAST Conference, Macau, July 10–14, 2001.

STURGEON, N. (1997). *Ecofeminist Natures: Race, Gender, Feminist Theory and Political Action.* New York and London. Routledge.

SWAIN, M. (2002). Gender/Tourism/Fun(?): An Introduction. In Swain, M. and Momsen, J. (eds.) *Gender/Tourism/Fun(?)* Elmsford, NY. Cognizant Communication Corporation: 1–14.

WALL, G. (1996). Ecotourism: Change, Impacts and Opportunities. Paper presented at the Ecotourism Equation: Measuring the Impacts Conference, Yale School of Forestry and Environmental Studies, April, 1996.

WALL, G. (2000). Ecotourism. In Jafari, J. (ed.) *Encyclopedia of Tourism.* New York and London. Routledge: 165–166.

Submitted: May 25, 2004
Accepted: July 27, 2004

Mass and urban ecotourism: new manifestions of an old concept

David B. Weaver

ABSTRACT

Ecotourism emerged as a nature-based manifestation of alternative tourism, which is still regarded by some academics and practitioners as the most legitimate manifestation of this sector. While an analysis of the three core criteria of ecotourism points to the suitability of high order protected areas as ecotourism venues, modified spaces also have potential with regard to the latter, and may have distinct environmental advantages. Urban places, in particular, accommodate at least five types of ecotourism product, suggesting that urban ecotourism is a complex and high carrying capacity phenomenon that can be focused around the themes of restoration and enhancement, while diverting some visitation away from vulnerable natural areas. The idea of mass ecotourism also remains somewhat contentious in light of ecotourism's origins, yet like urban ecotourism is an increasingly recognized phenomenon that has the potential to produce positive environmental impacts because of its economies of scale and tendency to be confined into manageable restricted spaces. However, for these advantages to be realized, it is essential that mass ecotourism be implemented in conformance with a comprehensive model of ecotourism that emphasizes ecosystems-based attractions, transformational education, and sustainable management that enhances the environment at a global scale.

Introduction

Ecotourism, like sustainable tourism and alternative tourism, is an example of an 'essentially contested concept', meaning that its use and application is inherently a matter of dispute and controversy (Gallie 1955/56, cited in Hall 1998). To many who study and otherwise participate in this burgeoning sector, 'proper' ecotourism is construed as a small-scale activity undertaken in relatively undisturbed natural sites within high order protected areas and other rural settings. This paper acknowledges and reinforces the legitimacy of this form of ecotourism, but argues that legitimate manifestations can also potentially occur at the other end of the continuum in highly modified urban settings, and as a form of mass tourism. Indeed, it is contended that such emerging non-traditional forms of ecotourism may even be more effective than its small-scale counterparts in realising the potential of this sector to generate positive environmental, socio-cultural and economic outcomes, if an appropriate philosophical and structural context is adopted.

Evolution of Ecotourism

The term 'eco-tourism' first appeared in the English language academic literature in the mid-1980s (Romeril 1985), although it is clear that tourism activity recognizable as such existed long before then (Hall 1998; Fennell 1999; Page and Dowling 2002). Nevertheless, the appearance of the term is significant and not merely of semantic interest, because it provided a catalyst around which formal investigation and development of the concept could subsequently be undertaken. These investigations have been greatly influenced by the seminal definition put forward by Ceballos-Lascurain in the late 1980s, which was disseminated widely in the pioneering ecotourism literature that followed (e.g. Ziffer 1989; Boo 1990):

> *ecotourism.... involves travelling to relatively undisturbed or uncontaminated natural areas with the specific object of studying, admiring, and enjoying the scenery and its wild plants and animals, as well as any existing cultural aspects (both past and present) found in these areas (Ceballos-Lascurain 1988: 13).*

His subsequent description of ecotourism or 'ecological tourism' as an activity undertaken in remote locations and involving 'modest' accommodation has also been influential, especially in popularising the view that it is a small-scale undertaking that contrasts with so-called 'mass' or high volume tourism. In fact, this perception mirrors the philosophical undercurrents that were then influencing many tourism academics and practitioners. Jafari (1989; 2001) describes the 1980s as the era of the 'adaptancy' platform, so-called because of the emergence of discussion about alternative forms of tourism purported to be better adapted than high volume tourism to the environmental and sociocultural circumstances of local communities and destinations. Essentially, ecotourism emerged as a nature based form of 'alternative tourism' whose ideal type was deliberately characterised by its opposition to unsustainable mass tourism, involving products and destinations that are small-scale (rather than the large-scale enterprises that characterise the mass tourism ideal type), rural (rather than urban), locally controlled (rather than controlled by external corporations), minimally mediated by the travel industry (rather than independently undertaken), etc. (Weaver 1993). This view of ecotourism as a more desirable alternative was widely supported, and continues to find strong support, in the literature (e.g. Johnston 1990; Fennell and Smale 1992; Prosser 1994; Sindiga 1999; Müller 2000).

Before examining whether ecotourism should be seen as a sector aligned exclusively with rural-focused alternative tourism, it is necessary to understand the three core criteria articulated by Blamey (1997; 2001) that most academics and practitioners now agree must be satisfied in order for an attraction or product to qualify as such:

- The attractions and products are primarily *nature based,* with associated cultural phenomena playing a supportive role. It is generally assumed that the associated plants and animals are non-captive, so that zoological parks, aquariums, botanical gardens and similar attractions are excluded;
- Qualifying products provide opportunities for *education and learning,* thereby distinguishing this sector from other forms of nature-based tourism in which the natural environment is basically a convenient setting for the pursuit of hedonistic or other non-educational motives; and
- These products and attractions are managed with the intent of being *environmentally, economically and socio-culturally sustainable,* which includes the involvement of and accrual of benefits to local communities.

Non-traditional Ecotourism Settings

These criteria point to the particular suitability of relatively isolated and undisturbed rural settings such as National Parks and other high order protected natural areas to serve as ecotourism venues. Such entities have a formal mandate to protect and preserve outstanding (i.e. attractive) natural resources, but in many cases are also required to accommodate complementary recreational activities that focus on the appreciation of these resources. Protected natural areas clearly should and do dominate the ecotourism literature (e.g. Ceballos-Lascurain 1996; Weaver 1998; Wearing and Neil 1999; Butler and Boyd 2000; Lawton 2001), but attention is beginning to be paid to the potential of other more modified sites to provide legitimate ecotourism experiences as well (Higham and Lück 2002). Lawton and Weaver (2001) and Weaver (2001a), for example, discuss the ecotourism capacities of non-traditional settings such as agricultural land, artificial reefs, abandoned transportation corridors (e.g., 'rails to trails' networks and defunct canals) and abandoned surface mines and quarries. Notably, they also consider the potential of sites within highly modified urban and ex-urban environments that are least likely to resonate with common sense.

The rationale for accepting and even encouraging ecotourism in these unconventional settings can be established by reconsidering the three core criteria of ecotourism within this context. First, with regard to nature-based attractions, congruency between these and modified settings is achieved if one accepts the caveat that the attractions of ecotourism can be based not only on the natural environment as a biome, but also on elements thereof, such as particular species of plant or animal (Weaver 2001a). Many native animals have become adapted to human modifications in the landscape, so that ecotourism opportunities are thereby created for example by the presence of migrating sandhill cranes in Nebraska grain fields, or by bats that have colonised an abandoned mine. On the education/learning dimension, the presence of native plants and animals in modified settings allows principles such as adaptability and co-existence, as well as habitat disappearance and rehabilitation, to be incorporated into product interpretation, thereby enriching the educational experience by emphasizing the complex interactions between the natural and cultural environment. Moreover, these experiences are available to a larger market because of the greater accessibility of modified spaces compared with relatively undisturbed environments.

Finally, with regard to the sustainability criteria, it can be argued that modified spaces are already substantially disturbed and, therefore, warrant a 'weak' approach to sustainable development conducive to either mass tourism or alternative tourism. This entails an anthropocentric and utilitarian perspective that encourages managed growth and exploitation within a conservationist framework that acknowledges the role of vital natural resources such as the ozone layer as 'critical capital' that cannot be replaced by a comparable amount of human capital (Hunter 1997). The accommodation and celebration of native plants and animals within altered environments is consistent with this approach, as is the possibility for habitat restoration and enhancement where appropriate and feasible. This weak approach also implies a high environmental (if not necessarily sociocultural) carrying capacity for ecotourism and ecotourists, since there are no fragile virgin habitats to take into account when weighing impacts. Perhaps more important is the role that modified spaces could play in diverting ecotourists from these vulnerable habitats, given the continuing rapid growth of the ecotourist market and the concomitant reduction in the area occupied by relatively undisturbed natural habitat. The residual insistence among some academics that ecotourism must occur in the spaces that are most likely to be negatively impact by this form of human activity is, indeed, a fundamental contradiction and, perhaps, a 'fatal flaw' of the definitions that still advocate this view.

Urban Ecotourism

Urban agglomerations are among the most modified environments, and hence invite particular scrutiny and skepticism as to their suitability for ecotourism. Clear advantages and disadvantages are immediately apparent. Among the former are proximity to services and markets (both residential and tourist), and diminished levels of concern that ecotourism will generate inadvertent negative environmental or socio-cultural consequences within an environmentally fragile setting. To the contrary, ecotourism product development in urban settings is far more likely to serve as a catalyst for environmental rehabilitation and other changes indicative of a sustainability model focused on enhancement rather than sustaining the status quo. Another opportunity in some urban areas is the oasis effect, wherein cities provide temporary or longer-term habitat for migratory and native animals that cannot find suitable habitat in adjacent intensive farmland. Urban residents, moreover, often enhance this role by actively feeding and providing shelter for native birds and other animals (Lepczyk *et al.* 2004). Among the obvious disadvantages are the hostility of most urban land uses and activities to most native plants and animals, and the presence of cultural and other 'built' forms of tourism that divert attention away from ecotourism. Another probable disadvantage is cognitive dissonance, wherein potential visitors, managers and others will see the idea of urban ecotourism as an oxymoron and will not be open to pursuing such opportunities. This may in turn be exacerbated by the tendency of some residents to regard certain wild animals (e.g., coyotes, raccoons, bats, deer) as opportunistic pests that interfere with and sometimes endanger humans. Of the more than 200,000 wildlife-human conflicts recorded by the US Department of Agriculture's Wildlife Service in 2003, 43 per cent occurred within an urban or suburban setting (Clayton 2004). Finally, urban environments offer only a semblance of any given natural ecosystem, which renders the interpretation experience incomplete.

The possibility of 'urban ecotourism' as a viable component of the urban tourism product has already been explicitly or implicitly cited in the literature. Higham and Lück (2002), for example, feature three successful case studies in New Zealand, including a penguin colony established in a harbour-side quarry, a dolphin-watching enterprise in Auckland's Hauraki Harbour, and a remnant natural habitat in the suburbs of Wellington. Notably, two of these case studies include a significant habitat rehabilitation component. The authors also mention an Auckland-based company that provides tours of volcanic and other geological sites within the city. Dodds and Joppe (2001) contribute to this nascent field with a 'green guide' to Toronto that includes ecotourism sites among a range of alternative urban tourist attractions.

The overall evidence, though incipient, anecdotal and fragmented, indicates the existence of an active urban ecotourism sector, though no attempts have yet been made to outline its structure. It is proposed here by way of a preliminary structure that urban ecotourism consists of five components, not necessarily mutually exclusive, arrayed along a continuum that more or less moves from a most 'unnatural' to a least 'unnatural' state:

1. *Structural sites* are drastically site-hardened locations, often in or near the central business district, that include buildings, roads and other land uses that offer minimal or no green space. Their overall ecotourism potential is limited, but does include a high degree of suitability for specialised activities such as the viewing of peregrine falcons that have been established in high-rise nesting sites in cities such as Pittsburgh, USA, and the viewing of bald eagles that have established nesting sites beneath bridges. Examples in Europe include medieval town centres where storks nest in the chimneys of some residences. Balconies, church spires and rooftop gardens offer other opportunities

for ecotourism. Illustrating the potential of bridges is the Congress Avenue Bridge in Austin, Texas, where a colony of 1.5 million Mexican free-tailed bats annually attracts an estimated 100,000 tourists and US$8 million in revenue. Interestingly, and perhaps as a portent for other potential urban wildlife attractions, the establishment of this successful urban ecotourism product was preceded by a period of widespread local hostility toward the bat colony based largely on stereotypical images and ignorance (Moreno 2004). The emerging rooftop garden and green roof movements also entail opportunities for urban ecotourism, since such facilities would provide a relatively safe microhabitat for birds, butterflies and other wildlife (Brenneisen 2004).

2. *Manicured green spaces* include highly manipulated sites such as municipal parks, university campuses, cemeteries and golf courses, as well as lawns, which provide habitat for a large number of native and exotic plants and animals. Many municipal parks already offer nature-based tours and trails, while Mount Auburn Cemetery in Cambridge, Massachusetts, has long been managed as a bird-watching site (Adams and Dove 1989). Golf courses have been widely condemned for their destructive environmental impacts, but the recent trend toward 'naturalistic' course design and management indicates a high potential to accommodate ecotourism (Terman 1997; Dodson 2000). Extensive fertilizer and pesticide application, necessary respectively to promote desirable vegetation and suppress undesirable vegetation, is an impediment for developing ecotourism opportunities on modified urban green spaces, as is the employment of exotic plant and grass species that displace native vegetation and have the potential to be invasive.
3. *Derelict and reclaimed sites* include quarries, rails-to-trails, land fill sites, relic industrial and military facilities, and other settings where the built urban landscape has been abandoned, creating wildlife habitat as well as possibilities for functional adaptation to ecotourism through rehabilitation and the establishment of supportive facilities and services. High land values and strategic locations often dictate that these newly available sites are used to build sports stadiums or luxury condominiums. However, the demand for quality urban green space is also extremely high in many cities, so that support for habitat rehabilitation and the creation of parkland is also usually strong. Toronto's Leslie Street Spit, a land fill site extending into Lake Ontario that is being restored as wildlife habitat, illustrates the ecotourism potential of derelict and reclaimed sites, and in particular indicates the potential of such sites to emphasize restoration and rehabilitation as a core component of interpretation and education. Another example is the Black Forest Urban Forest Project in the West Midlands of the UK, which includes the reforestation of highly contaminated former industrial land and waste sites (cited in Cloke and Jones 2000).
4. *Remnant 'natural' habitat* incorporates such areas as riparian buffer zones, shorelines, and remnant woodlot and wetland blocks or strips that resemble a natural environment, but do so only superficially because of the alterations caused by fragmentation, the invasion of exotic species, predation from domestic animals, exaggerated border effects, dumping of debris, and exposure to contaminated runoff and poor air quality. Such landscapes often exist within municipal parks or as a deliberate component of planned subdivisions (on homeowner or homeowner association property). As with derelict and abandoned settings, ecotourism offers opportunities for restoration and enhancement.
5. *Extensive natural habitat* is usually found on the urban periphery, often as a formal public protected area. While as such they constitute a traditional ecotourism venue, they are still significantly affected by this proximity to the city, in some cases because they are managed by municipal authorities vested with an urban mindset and a mandate to meet the needs of urban constituents. In most cases, such areas are additionally subject, at least at its edges, to the same forces that impoverish remnant natural habitats. More than any other category of urban ecotourism, extensive natural habitats are most likely to have been used for ecotourism prior to their incorporation into spreading urban environments, as illustrated by protected areas that are becoming assimilated into the suburbs of Sydney, Australia (Dwyer and Edwards 2000). An European variant of this model is the managed forest that surrounds major cities such as Paris and Berlin.

These five components suggest that urban ecotourism is extremely diverse and complex, particularly in large urban areas. This complexity is evident in the fact that many potential ecotourism sites combine the components or are undergoing transformation from one component to another. The latter scenario is illustrated by the Leslie Street Spit, which will eventually resemble an extensive natural area whose genesis as a derelict waste disposal site will then be irrelevant for all but interpretive purposes. Sewage lagoons, such as the Hornsby Bend site in Austin, Texas where 118 species of bird have been recorded (Bonta 1997), both combine and transform the categories, since they are built facilities, often located on derelict land, that can be groomed into a park-like condition or allowed to develop into a 'natural' wetland that include structural sites such as dams and sluices.

At a broader urban scale, relevant managerial considerations include the use of the corridor concept to link the network of spaces described above in order to reinforce and enhance the ecotourism potential of the whole urban ecotourism system of a particular city (Adams and Dove 1989). The diverse nature of such systems also points to the logic of collaborative partnerships involving municipal governments, community associations, non-profit organizations, educational institutions and the private sector. Visitation to these sites, more so than in rural areas, is likely to be dominated by local residents, so that there is a greater possibility of the tourism dimension being obscured or under-utilized. In this respect, the conventional tourism industry can play a crucial role by promoting these opportunities to hotel guests, convention attendees and other categories of urban tourist as a diversionary activity.

Mass Ecotourism

The potential involvement of community groups as well as the conventional tourism industry suggests that urban ecotourism can occur in the guise of either alternative tourism (for example, through neighbourhood associations conducting small group tours of suburban remnant riparian buffer zones) or mass tourism (as for example through a corporation's funding of a sophisticated high capacity interpretation centre built on the edge of a sewage lagoon). As with 'urban ecotourism', the concept of 'mass ecotourism' is still contested and contentious given the aforementioned genesis of this sector as a nature based form of alternative tourism, and the continuing tendency of many academics and practitioners to resist its affiliation with mass tourism. Clearly, there are situations, such as Antarctica, where it is most appropriately undertaken as alternative tourism (although the disingenuousness of the purist argument is demonstrated by the reliance of Antarctic visitors on major air carriers to access points of embarkation in Patagonia or Tasmania). However, as illustrated by the urban settings described above, there are other situations where ecotourism as mass tourism may not only be feasible, but desirable.

The idea that ecotourism can occur as mass tourism is actually quite established in the literature at least as an implicit construct. Already in the late 1980s, Laarman and Durst (1987) distinguished between 'hard' and 'soft' ecotourism, a contention that was subsequently echoed by Lindberg (1991), who differentiated between 'hard-core', 'dedicated', 'mainstream' and 'casual' ecotourists, and the government of Queensland (1997), which identified 'self-reliant', 'small group' and 'popular' modes of ecotourism. Extensive empirical research has corroborated this idea that ecotourism products can be arrayed along a continuum anchored by hard-core and soft-core ideal types (e.g. Palacio and McCool 1997; Weaver and Lawton 2002). Generally speaking, the hard core ideal type is characterized by small groups of ecotourists with strong environmentalist tendencies who seek mentally and physically challenging encounters with the natural environment in relatively remote settings where few if any services and facilities are available. In contrast, soft ecotourism is associated with large numbers of participants who seek short term, convenient ecotourism experiences as part of a multi-dimensional tourism trip mediated extensively by the tourism industry (Weaver 2001a). The actual words 'mass ecotourism' are still only rarely encountered, and then mainly as a pejorative term (Pleumarom 1993). As a term with positive connotations, the usage is more recent (e.g. Weaver 2001b).

As long as these 'soft', 'popular' or 'mass' experiences adhere to the three core criteria cited above, there is no inherent reason why this activity, involving tens of millions of participants each year in what is clearly a form of mass tourism, should be discounted as a legitimate manifestation of ecotourism. The argument against this link appears to be based in part on the strong likelihood of deleterious outcomes associated with high volumes of activity, and in part on concerns that the participants are not especially committed to the principles of sustainability or education. Accordingly, such products, even if they are not just 'pseudo-ecotourism' experiences that really do not adhere to the core criteria, can be easily transformed into the unsustainable type of mass tourism that will undermine the destination's environmental and socio-cultural integrity. This is a legitimate argument, but one that underestimates the potential advantages of mass ecotourism and overestimates the potential advantages of hard ecotourism. As pointed out by Weaver (2002), there are many more soft ecotourists than hard ecotourists to serve as potential agents of negative change, but the extent of these changes is limited by the tendency of soft ecotourists to congregate within relatively narrow spatial and temporal parameters, as for example in the vicinity of interpretation centres during the peak tourism season. While some view these small modified areas as 'sacrificial spaces' that spare other areas from undue tourist impacts, such confinement, at least in theory, produces economies of scale that justify sophisticated site-hardening and site-softening measures that mitigate the negative impacts of intensive use while potentially providing satisfying and effective interpretation and education. The long range plan for the intensively visited South Rim visitors' area of Grand Canyon National Park illustrates the possibilities (National Park Service 2001), as does other iconic National Parks such as Great Smokey Mountains (USA), Kruger (South Africa) and Lake District (UK). Conversely, hard ecotourists, though few in numbers, can inadvertently produce negative impacts in extensive areas designated as wilderness or natural. They have been associated, for example, with the transmission of contagious diseases to mountain gorillas in east Africa (Nowak 2001), and with the introduction of fecal bacteria in remote pools within Lamington National Park, Australia (Warnken and Buckley 2004). Their motives, moreover, can be based as much on egotism and elitism as environmentalism, according to Wheeller (1997).

There are also many more soft ecotourists than hard ecotourists to serve as potential agents of *positive* change. The main consideration here is ecotourism's alleged capacity for providing financial incentives to protect (in the case of high order protected areas) and restore (in the case of modified spaces) natural habitat (Lindberg 1998). High spending hard ecotourists may be willing to pay elevated user fees to access remote protected areas, but cumulative revenues from this small market may not be sufficient to justify the exclusion of extensive wild lands from less benign but more financially lucrative forms of commercial exploitation. The billions of dollars obtained directly and indirectly from soft ecotourists in places such as the Great Barrier Reef and the Great Smoky Mountains National Park, in contrast, better facilitate the incentive effect. More extensive benefits, in addition, may result if the mass ecotourism experience induces attitudinal and behavioural changes in the audience, including a heightened awareness of and responsiveness to environmental problems, through transformative interpretation (Christie and Mason 2003). Evidence for this effect is provided by Orams (1997), who found that visitors who participated in an interpretive programme at the Tangalooma dolphin observation facility in Queensland, Australia, were more likely to indicate that the experience made them more concerned about environmental issues.

Comprehensive Mass Ecotourism

Indeed, one of the greatest challenges in operationalizing sustainable mass ecotourism will be the development of interpretation and other educational strategies that satisfy mass tourists while inducing change in their attitudes and behaviour so that the negative impacts associated with high levels of visitation are minimized. To bring about the latter outcome, Weaver (in press) advocates the application of a 'comprehensive' model of engagement to the soft or mass ecotourism sector. In contrast to 'minimalist' ecotourism that focuses on particular charismatic megafauna and tries to sustain the status quo of the operation's site or locality, the comprehensive model adopts a holistic ecosystems approach to the attraction base, seeks transformative educational outcomes through 'deep' learning experiences, and supports the enhancement of the environment at a global scale. Mass ecotourism, if pursued only along a minimalist trajectory, is much more likely to evolve into the unsustainable mass tourism model feared by the skeptics. Comprehensive mass ecotourism, in contrast, harnesses the economies of scale discussed above to effect large-scale positive changes in the environment and in the attitudes of participants toward the environment. As such, the idea is only in its conceptual stage, though the experience of Grand Canyon National Park and other selected high order protected areas offer suggestions as to how the concept can be implemented. Non-traditional urban settings, where the minimalist approach might serve only to perpetuate an unsustainable status quo, offer intriguing possibilities for the implementation of the comprehensive mass ecotourism model.

Conclusion

Traditional thinking links ecotourism with rural based alternative tourism, and regards concepts such as urban ecotourism and mass ecotourism as oxymoronic and subversive. Yet, nothing innate to the core criteria of ecotourism indicates that this sector cannot be extended into these non-traditional frameworks. To do so not only dramatically extends the range, exposure and influence of ecotourism, but in theory provides the basis for a more sustainable model of the latter. Urban ecotourism, for example, is a complex system that can serve to divert visitation from sensitive natural environments to urban settings that are capable of accommodating large numbers of visitors. Moreover, the desire to increase the ecotourism potential of such areas lends itself to management and education strategies that emphasize habitat rehabilitation and enhancement, as per the comprehensive model of ecotourism. Mass ecotourism, whether applied to urban or rural venues, is more capable than alternative ecotourism of actualizing the incentive effect of tourism because of the economies of scale it generates, while confining direct impacts to restricted spaces where the latter can be mitigated by sophisticated management measures. It is imperative, in this regard, that advocacy for mass ecotourism be accompanied by efforts to pursue this along the comprehensive trajectory. Given the continuing rapid growth of the sector and its penetration into an array of environmentally and socio-culturally vulnerable settings, this may be the single most important management issue facing contemporary ecotourism.

References

ADAMS, L. and DOVE, L. (1989). *Wildlife Reserves and Corridors in the Urban Environment: A Guide to Ecological Landscape Planning and Resource Conservation.* Columbia MD. National Institute for Urban Wildlife.

BLAMEY, R. (1997). Ecotourism: The Search for an Operational Definition. *Journal of Sustainable Tourism* 5(2):109–130.

BLAMEY, R. (2001). Principles of Ecotourism. In Weaver, D. (ed.) *The Encyclopedia of Ecotourism.* Wallingford, UK. CABI Publishing: 5–22.

BONTA, M. (1997). Turning Wastewater into Wetlands. *Living Bird* 16(4): 22–26.

BOO, E. (1990). *Ecotourism: The Potentials and Pitfalls* Volume 1. Washington, DC. World Wildlife Fund.

BRENNEISEN, S. (2004). Green Roofs – How Nature Returns to The City. In Junge-Berberovic, R., Bächtiger, J. and Simpson, W. (eds.) *Proceedings of International Conference of Urban Horticulture.* Waedenswil, Switzerland. Acta Horticulturae: 289–293.

BUTLER, R.W. and BOYD, S. (eds.) (2000). *Tourism and National Parks: Issues and Implications.* Chichester, UK. Wiley.

CEBALLOS-LASCURAIN, H. (1996). Tourism, *Ecotourism, and Protected Areas.* Switzerland. World Conservation Union.

CEBALLOS-LASCURAIN, H. (1988). The Future of Ecotourism. *Mexico Journal* January 27: 13-14.

CHRISTIE, M. and MASON, P. (2003). Transformative Tour Guiding: Training Tour Guides to be Critically Reflective Practitioners. *Journal of Ecotourism* 2: 1–16.

CLAYTON, M. (2004). In Suburbia, It's a Wild, Wildlife. *Christian Science Monitor* September 16.

CLOKE, P. and JONES, O. (2000). From Wasteland to Woodland to Little Switzerland: Environmental and Recreational Management in Place, Culture and Time. In Font, X. and Tribe, J. (eds.) *Forest Tourism and Recreation: Case Studies in Environmental Management.* Wallingford, UK. CABI: 161–182.

DODDS, R. and JOPPE, M. (2001). Promoting Urban Green Tourism: The Development of the Other Map of Toronto. *Journal of Vacation Marketing* 7: 261–267.

DODSON, R. (2000). *Managing Wildlife Habitat on Golf Courses.* Ann Arbor, MI. Ann Arbor Press.

DWYER, L. and EDWARDS, D. (2000). Nature-Based Tourism on the Edge of Urban Development. *Journal of Sustainable Tourism* 8: 267-287.

FENNELL, D. (1999). *Ecotourism: An Introduction.* London. Routledge.

FENNELL, D. and SMALE, B. (1992). Ecotourism and Natural Resource Protection: Implications of an Alternative Form of Tourism for Host Nations. *Tourism Recreation Research* 17(1): 21–32.

GOVERNMENT OF QUEENSLAND (1997). *Queensland Ecotourism Plan.* Brisbane, Department of Tourism. Small Business and Industry.

HALL, C.M. (1998). Historical Antecedents of Sustainable Development and Ecotourism: New Labels on Old Bottles? In Hall, C.M. and Lew, A. (eds.) *Sustainable Tourism: A Geographical Perspective.* Harlow, UK. Longman: 13–24.

HIGHAM, J. and LÜCK, M. (2002). Urban Ecotourism: A Contradiction in Terms? *Journal of Ecotourism* 1: 36–51.

HUNTER, C. (1997). Sustainable Tourism as an Adaptive Paradigm. *Annals of Tourism Research* 24: 850–867.

JAFARI, J. (1989). An English Language Literature Review. In Bystrzanowski, J. (ed.) *Tourism as a Factor of Change: A Sociocultural Study.* Vienna. Centre for Research and Documentation in Social Sciences: 17–60.

JAFARI, J. (2001). The Scientification of Tourism. In Smith, V.L. and Brent, M. (eds.) *Hosts and Guests Revisited: Tourism Issues of the 21st Century.* New York. Cognizant: 28–41.

JOHNSTON, B. (1990). Introduction: Breaking Out of the Tourist Trap. *Cultural Survival Quarterly* 14(1): 2–5.

LAARMAN, J. and DURST, P. (1987). *Nature Travel and Tropical Forests.* Raleigh, NC. North Carolina State University.

LAWTON, L. (2001). Ecotourism in Public Protected Areas. In Weaver, D. (ed.) *The Encyclopedia of Ecotourism.* Wallingford, UK. CABI Publishing: 287–302.

LAWTON, L. and WEAVER, D. (2001). Modified Spaces. In Weaver, D. (ed.) *The Encyclopedia of Ecotourism.* Wallingford, UK. CABI Publishing: 315–326.

LEPCZYK, C., MERTIG, A. and LIU, J. (2004). Assessing Landowner Activities Related to Birds Across Rural-to-Urban Landscapes. *Environmental Management* 33: 110–125.

LINDBERG, K. (1991). *Policies for Maximizing Nature Tourism's Ecological and Economic Benefits.* Washington, DC. World Resources Institute.

LINDBERG, K. (1998). Economic Aspects of Ecotourism. In Lindberg, K., Wood, M.E. and Engeldrum, D. (eds.) *Ecotourism: A Guide for Planners and Managers.* Volume 2. Washington, DC. The Ecotourism Society: 87–117.

MORENO, S. (2004). City Embraces the Bats Who Came Home to Roost. *Washington* Post July 19: A3.

MÜLLER, F. (2000). Ecotourism: An Economic Concept for Ecological Sustainable Tourism. *International Journal of Environmental Studies* 57: 241–251.

NATIONAL PARK SERVICE (2001). *Canyon View Information Plaza.* http://www.nps.gov/grca/publications/cvip.pdf. Accessed on 13 July 2004.

NOWAK, R. (2001). Tourists in the Mist Endanger Gorillas. *New Scientist* 169: 18.

ORAMS, M. (1997). The Effectiveness of Environmental Education: Can We Turn Tourists into 'Greenies'? *Progress in Tourism and Hospitality Research* 3: 295–306.

PAGE, S. and DOWLING, R. (2002). *Ecotourism.* London. Prentice-Hall.

PALACIO, V. and MCCOOL, S. (1997). Identifying Ecotourists in Belize Through Benefit Segmentation: A Preliminary Analysis. *Journal of Sustainable Tourism* 5: 234–243.

PLEUMAROM, A. (1993). What's Wrong with Mass Ecotourism. *Contour* 6(3/4): 15–21.

PROSSER, R. (1994). Societal Change and the Growth in Alternative Tourism. In Cater, E. and Lowman, G. (eds.) *Ecotourism: A Sustainable Option?* Chichester, UK. Wiley: 19–37.

ROMERIL, M. (1985). Tourism and the Environment – Towards a Symbiotic Relationship. *International Journal of Environmental Studies* 25: 215–218.

SINDIGA, I. (1999). Alternative Tourism and Sustainable Development in Kenya. *Journal of Sustainable Tourism 7:* 108–127.

TERMAN, M. (1997). Natural Links: Naturalistic Golf Courses as Wildlife Habitat. *Landscape and Urban Planning* 38: 183–197.

WARNKEN, W. and BUCKLEY, R. (2004). Instream Bacteria as a Low-Threshold Management Indicator of Tourist Impacts in Conservation Reserves. In Buckley, R. (ed.) *Environmental Impacts of Ecotourism.* Wallingford, UK. CABI Publishing: 325–337.

WEARING, S. and NEIL, J. (1999). *Ecotourism: Impacts, Potentials and Possibilities.* London. Butterworth-Heinemann.

WEAVER, D. (1993). Ecotourism in the Small Island Caribbean. *GeoJournal* 31: 457–465.

WEAVER, D. (1998). *Ecotourism in the Less Developed World.* Wallingford, UK. CAB International.

WEAVER, D. (2001a). *Ecotourism.* Brisbane. John Wiley and Sons.

WEAVER, D. (2001b). Ecotourism as Mass Tourism: Contradiction or Reality? *Cornell Hotel and Restaurant Administration Quarterly* 42(2): 104–112.

WEAVER, D. (2002). The Evolving Concept of Ecotourism and its Potential Impacts. *International Journal of Sustainable Development* 5: 251–264.

WEAVER, D. (in press). Comprehensive and Minimalist Dimensions of Ecotourism. *Annals of Tourism Research.*

WEAVER, D. and LAWTON, L. (2002). Overnight Ecotourist Market Segmentation in the Gold Coast Hinterland of Australia. *Journal of Travel Research* 40: 270–280.

WHEELLER, B. (1997). Here We Go, Here We Go, Here We Go Eco. In Stabler, M. (ed.) *Tourism & Sustainability: Principles to Practice.* Wallingford, UK. CAB International: 39–50.

ZIFFER, K. (1989). *Ecotourism: The Uneasy Alliance.* Washington, DC. Conservation International.

Submitted: July 13, 2004
Accepted: October 21, 2004

Examining the relationship between ecotourists and philanthropic behaviour

Martha L. Barnes and Paul Eagles

ABSTRACT

Ecotourism has some of its programmes and initiatives influenced by the philanthropic behaviour of its constituents. Philanthropic behaviour is demonstrated through the contributions of time or charitable donations. In Kenya, philanthropic behaviour has been instrumental in generating revenue for the Kenya Wildlife Service. Conservationists see ecotourism as a strategy for providing funds for conservation while at the same time justifying its importance. In Canada, the United States of America, Britain and Australia, individuals have donated CDN $5 billion, US $135 billion, £7 billion and A$3 billion, respectively. Furthermore, Canadians volunteered 1 billion hours in 2000, Americans volunteered 199 billion hours in 1998 and Australians volunteered 704 million hours. Some of this donated money and time was influenced by and directed to ecotourism activities.

Introduction

Philanthropic behaviour occurs in the world of ecotourism and it is crucial to examine possible factors influencing this behaviour. The philanthropic behaviour by ecotourists has not been adequately addressed in the literature. For instance, the literature is unable to answer a question such as 'what motivates ecotourists to engage in philanthropic behaviour?' There is some literature which suggests that individuals donate to organizations that represent values and beliefs that mirror their internal belief system (Martin 2000). A similar explanation may exist for ecotourists. Research also supports the notion that past behaviour is a strong predictor of future behaviour (Martin 2000), thereby suggesting that ecotourists that have engaged in philanthropic behaviour in the past are more likely to engage in future philanthropic behaviour. And finally, research suggests that when individuals visit, experience, and appreciate a place, they are more likely to make a charitable donation (Butler 1992), meaning that all ecotourists have the potential to help preserve areas they have experienced. Butler believes that for an area to be managed and protected, constituents must exist, and in the case of ecotourism, the constituents are ecotourists. The intent of this discussion paper is to make a call for research in understanding the link between ecotourists and philanthropic behaviour. Clearly, not all ecotourists engage in philanthropic acts. The extent to which philanthropic behaviour is pursued by ecotourists is unknown and so until such time that we can say why ecotourists engage in philanthropic behaviour, we are left making generalizations in the hopes of spurning some future research.

Definition of Key Terms

Philanthropic Behaviour: Philanthropic behaviour refers to all activity which promotes human well-being including blood donations, gifts in kind, voluntary work, and financial contributions (Reddy 1980). In Canada, the year 2000 saw approximately 22 million individual Canadians contribute over CDN $5 billion (Hall *et al.* 2001) to a wide spectrum of charitable organizations. In Britain, even though researchers suggest that there is a slow-down in charitable giving, between 2001 and 2002 a total of £10 billion was generated (Greene 2003). In the United States of America, annual individual charitable contributions increased from US $59 billion in 1964 to US $105 billion in 1994 (Kaplan and Ekstrom 1994). Though research is just beginning to be collected on philanthropy in Australia, in 1997 Australians donated A$3 billion (Lyons and Hocking 2000).

Furthermore, the number of volunteered hours also accounts for a demonstration of the 'philanthropic' nature of individuals. For instance, in Canada individuals volunteered over 1 billion hours in 2000 (Hall *et al.* 2001). A national study in the United States found that 109 million individuals volunteered 199 billion hours of time in 1998

(Wiener *et al.* 2002). In Australia, the year 2000 saw 4 million volunteers for a total of 704 million hours, excluding the Sydney Olympics (Lyons and Hocking 2000). It is important to examine both volunteering and charitable giving because these two acts often overlap. Individuals who volunteer are more likely to also make charitable donations (Mixer 1993). In Canada, 93% of individuals who volunteered at least 188 hours or more (according to the categorization identified by Hall *et al.*) annually also made charitable contributions (Hall *et al.* 2001). Individuals who volunteered less than 188 hours did not make as many charitable donations.

Prosocial Behaviour: A concept that is similar to philanthropic behaviour, prosocial behaviour refers to 'any behaviour which has positive outcomes for individuals' (Cadenhead and Richman 1996: 171). While this term is sometimes used interchangeably with philanthropic behaviour, prosocial behaviour seems to be a more theoretical construct. Nonetheless, understanding prosocial behaviour contributes to understanding philanthropic behaviour.

Ecotourism: Nature-based, alternative, nature-oriented, natural or green tourism are terms often used to describe the concept encompassing 'ecotourism'. Dowling (2000:11) defines ecotourism as 'special interest tourism which places emphasis on greater contact and understanding between hosts and guests as well as between tourists and the environment'. Dowling also identifies several components of true ecotourism that include ecologically sustainable activities, conservation, active interpretation, education, and support for local economy.

Theoretical Explanations for Philanthropic Behaviour

Philanthropic behaviour can be explained by borrowing from Bar-tel (1976) and Blau (1964). More specifically, an exchange approach states that individuals perform acts for reasons related to future rewards (Bar-tel 1976). Exchange theorists consider pure acts of altruism rare (Blau 1964) and that all acts are performed with the expectation of some reward, whether it be tangible or intangible. According to Martin (2000) when individuals benefit from a particular service or organization (or are likely to in the future) they are more likely to make a donation. Therefore, it is possible that individuals who travel to an ecotourism destination and benefit (emotionally, socially, and financially), may feel that their donation may help preserve the destination so that it is available for a future visit.

Another explanation put forth by Bar-tel (1976) to understand the phenomena of prosocial behaviour which can be applied to the ecotourism-philanthropic phenomena is a developmental approach. In other words, by exploring the role that socialization plays in motivating individuals to act in prosocial ways, researchers acknowledge that philanthropic behaviour is a learned behaviour. Smith and Bakdwin (1974) gathered data with adult volunteers that demonstrated a substantial portion of the variance (10%- 25%) in an individual's formal voluntary organization participation was accounted for by the joint impact of four types of indices: parental socio-economic status, the volunteers' socio-economic status, parental formal voluntary organization attitudes, and parental formal voluntary organization behaviour (in other words, being socialized to give). These findings suggest that individuals who are exposed to prosocial behaviour as children are more inclined to engage in prosocial behaviour as an adult. More recently, research has found a positive relationship between respondents' volunteering and parent behaviour (Hall *et al.* 2001). Specific to the topic of ecotourism, it could be that ecotourism is structured in a way that philanthropic behaviour is part of the culture and individuals are socialized to follow that lead. The two theoretical explanations offered provide insights into the reasons that individuals engage in philanthropic behaviour; however, research has yet to be done in an ecotourism setting.

Evidence of Philanthropic Pursuits by Ecotourists

This paper will now present some of the successful strategies that have been employed to strengthen the natural resources management that connects ecotourists with philanthropy. There has been a steady increase in the number of tourism corporations encouraging philanthropic behaviour by its visitors. Ecotour companies have been known to develop programmes that encourage their clients to donate money towards conservation causes at the ecotourist destinations. For example, Abercrombie and Kent, a British-based tour operator, established a non-profit conservation organization to provide financial support to numerous protected areas in Kenya. As part of a tour package, tourists are taken to protected areas sponsored by Abercrombie and Kent to increase financial support (Brandon 1996). Another example is found in the Galapagos Islands where Lindblad Expeditions created a successful customer-generated charitable fund supporting scientific research and environmental preservation efforts (Ham 2001). Beginning in 1997, this project collected approximately US $4,000 a week ($500,000 total) from tourists' donations. The money funded such programmes as the eradication of feral pigs from the island of Santiago, the operation of a patrol boat to combat illegal fishing, and the establishment of local environmental education (Ham 2001).

Generally, ecotourists are more interested in environmentally-appreciated trips than environmentally-consumptive sport hunting trips (Boo 1990). Ecotourists often have a strong commitment to the environment and to its conservation (Eagles 1992; Wearing and Neil 1999). Furthermore, when individuals travel to undeveloped countries, they may feel inclined to help host communities. For example, ecotourists are 8.5% more likely than regular tourists to purchase products produced by local, environmentally responsible individuals while vacationing (Wearing and Neil 1999). Furthermore, since ecotourists like to witness their expenditures resulting in conservation, many will contribute to local conservation organizations in the hope of furthering education and preservation of that particular area (Eagles and Higgins 1998). In the past, this concern for the environment and its preservation has led to widespread public pressure to conserve resources through the establishment of parks and protected areas (Boo 1990). In other research, this fortuitous connection between travel, traveller attitudes, public policy and conservation is known as the Tourism-Conservation-Attitude Cycle (Eagles and McCool 2002). Fundamental to this cycle is the philanthropic behaviour of travellers (i.e. ecotourists).

In terms of the motives associated with ecotourists, these individuals are often motivated to visit a destination classified as a national park or some other form of protected area. For instance, Obua and Harding (1996) found that 35% of ecotourists reported 'status as a protected area' as a motive for travel. Research has found that Canadian ecotourists were motivated by the opportunity to experience nature and natural surroundings, learn about nature, be active and connect with individuals of similar interests. Furthermore, Canadians were motivated to travel to see tropical forests, wilderness and undisturbed nature as well as learn about nature. Each of these motives was often experienced in a protected area (Eagles 1992).

Ecotourists' conservation motives are often directed beyond the protection of special sites to the conservation of wildlife species. A widely-discussed connection between travel motives and government policy is that of whale watching and whale conservation. Increasingly restrictive whale exploitation polices are correlated with vast increases in whale watching and whale appreciation. By heightening the public's awareness of the whale crisis, a strong voice was heard and government responded positively (Butler 1992). A study investigating the whale watching industry in Tonga, a small island in the South Pacific, found evidence supporting the ecotourism-whale conservation argument (Orams 2001). Starting in 1978, whale hunting was banned in Tonga. In recent years there has been pressure by some local people and the World Council of Whalers Association to allow whale hunting once again, much to the dismay of tourism and environmental groups. Orams found that 83% of individuals arriving by yacht and 95% of individuals arriving by plane to the island were opposed to the commencement of commercial whaling. More importantly, when asked if whale hunting would reduce their likelihood of visiting any island in particular, 62% of yacht visitors and 78% of airplane visitors responded "yes". When asked if whale hunting would affect their specific travel plans to Tonga, 65% of yacht visitors and 73% of aircraft visitors said they would be less inclined to visit. In other words, individual travellers wish to protect whales and make their travel choices based on these conservation principles.

In Uganda, as a result of the work of Dian Fossey, a non-profit foundation was established to accept funds for research and conservation. Gorilla trekking in Bwindi National Park in Uganda was part of an estimated US $1.5 million donor funded project being used to promote tourism in the forest. In Rwanda in 1991, before the civil war, 75% of the tourism income was generated from visitation of the gorillas, equalling US $1 million in gate fees and US $9 million in indirect expenditures (Brandon 1996). A portion of this revenue went directly to the Mountain Gorilla Project and Dian Fossey Gorilla Fund. The connection between ecotourism, conservation, and philanthropic behaviour is apparent in this eastern Africa gorilla-park-ecotourism activity.

Nianyong and Zhuge (2001) observed that in developed and developing countries, industry and private donations are important funding sources. A non-profit organization is often instrumental to many conservation partnerships because in many Third World countries, non-profit organizations act as conduits for funds from foreign aid agencies and private donors. An advantage of non-profit organizations is that they typically have long-term sustainability in mind. According to Selin (1999), the majority of tourism partnerships are cross-sector initiatives that involve individuals from industry, government and voluntary organizations. For instance, the World Wide Fund for Nature Arctic project includes conservationists, tourism academics, researchers, local and national government and Arctic community representatives all working together to integrate conservation ideals into tourism projects (Mason, Johnston and Twynam 1999). In addition to individual monies, granting agencies and partnerships, international donors often form partnerships with local agencies as a way of contributing to the environment. For example, USAID promotes a tourism strategy called 'Low Impact Tourism'. What is interesting about this project is that the entire plan is funded through private donations and investments (Brandon 1996).

Conclusion

In summary, this discussion paper has highlighted some examples that exist in the field of ecotourism where philanthropic behaviour is apparent. This theoretical structure of possible explanations of philanthropic behaviour by ecotourists merits further empirical and scholarly inquiry. The reasons that ecotourists are inclined to preserve wildlife, contribute to local economy, lend political influence, and volunteer must be further uncovered. There are several possible explanations for this philanthropic behaviour that need to be verified in the field. Insights as to the reasons for philanthropic behaviour are necessary to ensure that future ecotourists remain committed both financially and emotionally to specific sites and that natural resources are used sustainably. The philanthropic tendencies of ecotourists may be one of the most important forces for conservation available.

References

BAR-TEL, D. (1976). *Prosocial Behaviour: Theory and Research.* Washington, D.C. John Wiley & Sons.

BLAU, P. (1964). *Exchange and Power in Social Life.* New York. John Wiley.

BOO, E. (1990). *Ecotourism: The Potentials and Pitfalls* (Vol. 1). Washington, D.C. World Wildlife Fund.

BRANDON, K. (1996). *Ecotourism and Conservation: A Review of Key Issues* (1st ed.). Washington, D.C. The World Bank.

BUTLER, J. R. (1992). Ecotourism: Its Changing Face and Evolving Philosophy. Paper presented at the IVth World Congress on National Parks and Protected Areas.

CADENHEAD, A. and RICHMAN, C. (1996). The Effects of Interpersonal Trust and Group Status on Prosocial and Aggressive Behaviors. *Social Behavior and Personality* 24(2): 169–184.

DOWLING, R. K. (2000). Global Ecotourism at the Start of the New Millennium. *World Leisure 2:* 11–19.

EAGLES, P. F. J. (1992). The Travel Motivations of Canadian Ecotourists. *Journal of Travel Research* 2: 3–7.

EAGLES, P. F. J. and HIGGINS, B. R. (1998). Ecotourism Market and Industry Structure. In Lindberg, K., Epler-Wood M. and Engledrum D. (eds.) *Ecotourism: A Guide for Planners and Managers* (2nd ed). North Bennington, VT. The Ecotourism Society: 11–43

EAGLES, P. F. J. and McCOOL, S. F. (2002). *Tourism in National Parks and Protected areas: Planning and Management.* Wallingford, UK. CABI.

GREENE, S. (2003). Britain's Largest Charities Face Slowdown in Private Donations. *Chronicle of Philanthropy* 15(2): 11–14.

HALL, M., MCKEOWN, L. and ROBERTS, K. (2001). *Caring Canadians, Involved Canadians: Highlights from the 2000 National Survey of Giving, Volunteering and Participating.* Ottawa, ON. Ministry of Industry.

HAM, S. (2001). A Theory-based Approach to Campaign Planning for Traveller's Philanthropy. Paper presented at the BEST Summit on Philanthropy, Nov. 9–11, Punta Cana, Dominican Republic.

KAPLAN, A. and EKSTROM, H. (1994). *Giving USA: The Annual Report on Philanthropy.* New York. American Association of Fund Raising Council.

LYONS, M. and HOCKING, S. (2000). *Dimensions of Australia's Third Sector.* Sydney, Australia. Centre for Australian Community Organizations and Management, University of Sydney.

MARTIN, S. (2000). Donations as an Alternative to Wilderness User Fees: The Case of the Desolation Wilderness. Paper presented at the USDA Forest Service Conference. May 22–27, Missoula, MT.

MASON, P., JOHNSTON, M. and TWYNAM, D. (1999). The World Wide Fund for Nature Arctic Tourism Project. *Journal of Sustainable Tourism* 8(4): 305–323.

MIXER, J. (1993). *Principles of Professional Fundraising.* San Francisco, CA. Jossey-Bass.

NIANYONG, H. and ZHUGE, R. (2001). Ecotourism in China's Nature Reserves: Opportunities and Challenges. *Journal of Sustainable Tourism* 9(3): 228–242.

OBUA, J. and HARDING, D. (1996). Visitor characteristics and Attitudes Towards Kibale National Park, Uganda. *Tourism Management* 17(7): 495–505.

ORAMS, M. (2001). From Whale Hunting to Whale Watching in Tonga: A Sustainable Future? *Journal of Sustainable Tourism* 9(2): 128–146.

REDDY, R. (1980). Individual Philanthropy and Giving Behavior. In Horton-Smith D., Macaulay, J. & Associates (eds.) *Participation in Social and Political Activities.* San Francisco, CA. Jossey-Bass: 370-399.

SELIN, S. (1999). Developing a Typology of Sustainable Tourism Partnerships. *Journal of Sustainable Tourism* 7(3 & 4): 260–273.

SMITH, D. H. and BAKDWIN, B.R. (1974). Parental Socialization, Socio-economic Status, and Volunteer Organization Participation. *Journal of Voluntary Action Research* 3(3&4): 59–66.

WEARING, S. and NEIL, J. (1999). *Ecotourists: Impacts, Potentials and Possibilities.* Oxford. Butterworth-Heinemann.

WIENER, S., TOPPE, C., JALANDONI, N., KIRSCH, A. and WEITZMAN, M. (2002). *Giving and Volunteering in the United States: Findings from a National Survey.* Washington, DC. Independent Sector.

Submitted: February 10, 2004
Accepted: May 12, 2004

Index

Note: **Bold** page numbers refer to tables and *italic* page numbers refer to figures.